Springer Tracts on Transportation and Traffic

Volume 14

Series editor

Roger P. Roess, New York University Polytechnic School of Engineering,
New York, USA
e-mail: rpr246@nyu.edu

About this Series

The book series "Springer Tracts on Transportation and Traffic" (STTT) publishes current and historical insights and new developments in the fields of Transportation and Traffic research. The intent is to cover all the technical contents, applications, and multidisciplinary aspects of Transportation and Traffic, as well as the methodologies behind them. The objective of the book series is to publish monographs, handbooks, selected contributions from specialized conferences and workshops, and textbooks, rapidly and informally but with a high quality. The STTT book series is intended to cover both the state-of-the-art and recent developments, hence leading to deeper insight and understanding in Transportation and Traffic Engineering. The series provides valuable references for researchers, engineering practitioners, graduate students and communicates new findings to a large interdisciplinary audience.

More information about this series at http://www.springer.com/series/11059

Andrzej Kobryń

Transition Curves for Highway Geometric Design

 Springer

Andrzej Kobryń
Faculty of Civil and Environmental
    Engineering
Bialystok University of Technology
Białystok
Poland

ISSN 2194-8119 ISSN 2194-8127 (electronic)
Springer Tracts on Transportation and Traffic
ISBN 978-3-319-53726-9 ISBN 978-3-319-53727-6 (eBook)
DOI 10.1007/978-3-319-53727-6

Library of Congress Control Number: 2017931572

Printed on acid-free paper

This Springer imprint is published by Springer Nature
The registered company is Springer International Publishing AG
The registered company address is: Gewerbestrasse 11, 6330 Cham, Switzerland

Contents

Chapter 1
Introduction

A designing of highways involves establishing the design details of the selected route, including final horizontal and vertical alignments, drainage facilities, and all items of construction. The design process of a highway involves preliminary location study, environmental impact evaluation, and final design. This process normally relies on a team of professionals, including engineers, planners, economists, sociologists, ecologists, and lawyers. Such a team may have responsibility for addressing social, environmental, land-use, and community issues associated with highway development. An important part of the highway design is a geometrical design.

The geometric design of roads and highways is a very complicated engineering task. It requires very often to take into account different terrain limitations, especially in mountainous and densely built areas (Figs. 1.1 and 1.2). Correct route design is associated with the use of appropriate optimization methods and the use of appropriate geometric elements that make it easy to adjust the route to those limitations.

The basic elements of geometric design of highways are horizontal alignment, vertical alignment, cross section, and intersection (Easa 2003). The horizontal alignment consists of straight sections (i.e. tangents) connected by horizontal curves, which are normally circular curves with or without transition curves. The basic design features of horizontal alignment include minimum radius, transition curves, superelevation and sight distance. The vertical alignment consists of straight roadway sections (grades or tangents) connected by vertical curves. The grade line is laid out in the preliminary location study to reduce the amount of earthwork and to satisfy other constraints such as minimum and maximum grades.

© Springer International Publishing AG 2017

A. Kobryń, *Transition Curves for Highway Geometric Design*, Springer Tracts on Transportation and Traffic 14, DOI 10.1007/978-3-319-53727-6_1

1

Fig. 1.1 Route in the mountainous terrain

Fig. 1.2 Route in the mountainous and densely built area

Two basic curves are used for connecting straight roadway sections in geometric design (Meyer and Gibson 1980, Lamm et al. 1999, Easa 2003, Rogers 2008, Brockenbrough 2009, Wolhuter 2015):

- a simple circular curve for horizontal alignment (Fig. 1.3)

  and

- a simple parabolic curve for vertical alignment (Fig. 1.4).

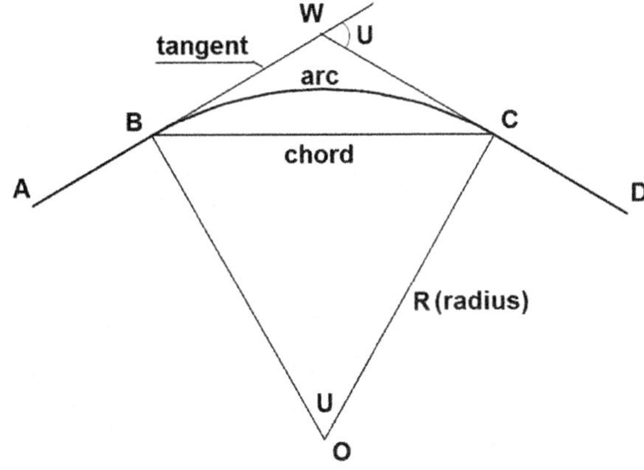

Fig. 1.3 Simple *circular curve* for *horizontal* alignment

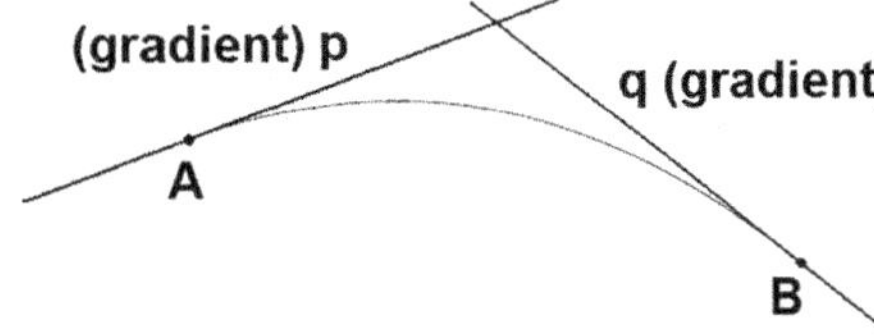

Fig. 1.4 Simple *parabolic curve* for *vertical* alignment

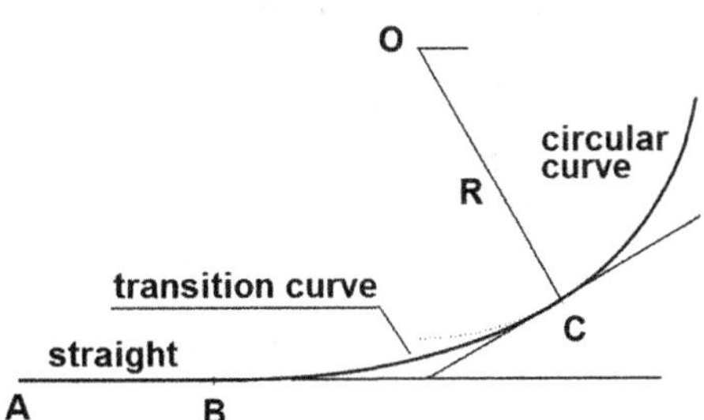

Fig. 1.5 Transition curve

Other options include transition curves (Fig. 1.5), compound curves, reverse curves and combined curves for horizontal alignment. In the case of vertical alignment possible options include unsymmetrical crest curves (vertical curves where the total change in gradient is negative) and sag curves (vertical curves where the total change in gradient is positive).

Transition curves fulfill a special role in the highway design. These types of curves are used to connect curved and straight sections of highway. They can also be used to make easier the change between two circular curves where the difference in radius is large. The purpose of transition curves is to permit the gradual introduction of centrifugal forces. The radius of curvature of a transition curve gradually decreases from infinity at the intersection of the tangent and the transition curve to the designated radius at the intersection of the transition curve with the circular curve.

A horizontal transition curve is a curve which radius continuously changes. It provides a transition between a tangent and a circular curve (simple transition curve) or between two circular curves with different radii (segmental transition curve). For simple transition curves, the radius varies from infinity at the tangent end to the radius of the circular curve at the curve end. For segmental transition curves, the radius varies from that of the first circular curve to that of the second circular curve.

The objectives for providing a horizontal transition curve are given below:

- to introduce gradually the centrifugal force between the tangent point and the beginning of the circular curve, avoiding sudden jerk on the vehicle. This increases the comfort of passengers.
- to enable the driver to turn the steering wheel gradually for his own comfort and security,
- to provide gradual introduction of super elevation, and
- to provide gradual introduction of extra widening.
- to enhance the aesthetic appearance of the road.

Vertical curves are traditionally designed as parabolic curves that are connected directly to the tangents (without transitions). A vertical transition curve has been developed for use before and after a parabolic curve (Easa and Hassan 2000a, b). The vertical transition curve consists of transition–parabolic–transition segments. Similar to the horizontal transition curve, the vertical transition curve is especially useful for sharp vertical alignments.

Different, classical types of transition curves are spiral or clothoid (spiral curve), cubic parabola and lemniscate. The spiral curve is recommended as the transition curve because it fulfills the requirements of an ideal transition curve, that are:

- rate of change or centrifugal acceleration is consistent (smooth) and

- radius changes linear at the any curve point.

Also other solutions of this type of curves were presented in the literature. Some of these curves will be presented in subsequent chapters of this book. A practical interesting group of curves are so-called polynomial transition curves. The purpose of this book is to present different transition curves in an orderly manner. In the first place it includes the presentation of possible ways to mathematical description of transition curves.

Different solutions of transition curves will be subsequently presented. Some of these curves have a geometry which is characteristic for the conventionally understood transition curves. In many cases, the nature of these solutions is more general than the classically understood transition curves. This approach bases on a broader definition of transition curves. The transition curves will be understood as such curves that connect any two points with the specified directions of tangents and radii of curvature. So understood transition curves can be used for routing of various counterparts geometric systems, which are listed earlier in this chapter. Their major advantage is that the entire geometrical transition between any points can be described by a single equation.

The book presents purely geometrical aspects related to the design described transition curves. In the book other aspects, such as dynamic aspects, shaping of superelevation, sight distance analysis or coordination of the horizontal and vertical alignment are omitted. Depending on the area of application of curves presented in the following sections of this work, these questions require separate analysis and research.

References

Brockenbrough RL (ed) (2009) Highway engineering handbook, 3rd edn. McGraw-Hill, Professional Book Group, New York

Easa SM, Hassan Y (2000a) Development of transitioned vertical curve. I Prop Transp Res Part A 34(6):481–486

Easa SM, Hassan Y (2000b) Development of transitioned vertical curve. II. Sight distance. Transp Res Part A 34(7):565–584

Easa SM (2003) Geometric design. In: The civil engineering handbook. Chen W.F, Liew J.Y.R, (ed), CRC Press, Taylor and Francis Group, Boca Raton

Lamm R, Psarianos B, Mailänder T (1999) Highway design and traffic safety engineering handbook. McGraw-Hill, Professional Book Group, New York

Meyer CF, Gibson DW (1980) Route surveying and design. Harper and Row, New York

Rogers M (2008) Highway engineering, 2nd edn. Wiley-Blackwell, Chichester-Oxford

Wolhuter KM (2015) Geometric design of roads handbook. CRC Press, Taylor and Francis Group, Boca Raton

Chapter 2
Simple Horizontal and Vertical Curves

In the geometric design of highways, circular curves as horizontal curves (Sect. 2.1) and parabolic curves as vertical curves (Sect. 2.2) are the most widespread. Apart from this type of curves, so-called transition curves are traditionally used as a geometric elements between the straight and the circular arc or between two circular arcs with different radii. The most popular transition curve is a clothoid (also known as spiral curve). Apart from the clothoid other solutions of transition curves are also known. They will be presented in the following sections of this work.

2.1 Circular Horizontal Curve

In the geometric design of horizontal curves a circular curves are very widespread (Brockenbrough 2009, Easa 2003, Lamm et al. 1999, Meyer and Gibson 1980, Rogers 2008, Wolhuter 2015). Figure 2.1 shows a circular curve (with radius R and centre O) joining two straights P'P and K'K with intersect at point W, where:

P and K   tangent points,
U angle of intersection of straights P'P and K'K.

The individual geometric elements occupy the following location:

- point S is the mid-point of the circular arc and the mid-point of the tangent line LM,
- point S lies on the line OW,
- Q is the mid-point of the chord PK and lies on the line OW,
- radii OA and OB intersect the straights P'P and K'K at right angles,

© Springer International Publishing AG 2017

A. Kobryń, *Transition Curves for Highway Geometric Design*, Springer Tracts on Transportation and Traffic 14, DOI 10.1007/978-3-319-53727-6_2

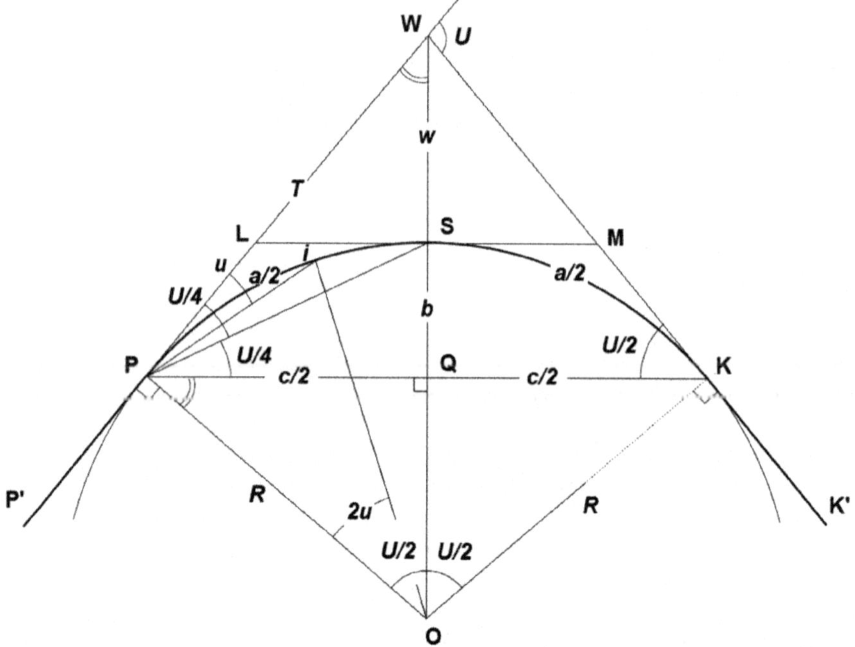

Fig. 2.1 Circular curve

- tangent line LM and the chord PK are parallel,
- the chord PK is perpendicular to the straight line OW.

  The following formulae may be deduced from Fig. 2.1:

- tangent length (PW = WK)

$$T = R \tan\frac{U}{2} \tag{2.1}$$

- arc length ($\overline{PK}$)

$$a = R \cdot U \tag{2.2}$$

- chord length (PK)

$$c = 2R \sin\frac{U}{2} \tag{2.3}$$

- mid ordinate distance (QS)

$$b = R\left(1 - \cos\frac{U}{2}\right) \tag{2.4}$$

- secant distance (SW)

$$w = R\left(\sec\frac{U}{2} - 1\right) \tag{2.5}$$

Circular curves can be used not only to design curvilinear transitions between two straights, as shown in Fig. 2.1. They can also be used for designing complex geometric systems, such as: compound circular curves, reverse circular curves and combined curves. Such systems should be understood as follows:

- compound circular curves—two or more consecutive circular curves with different radii (Fig. 2.2),
- reverse circular curves—two or more consecutive circular curves, with the same or different radii which centres lie on different sides of a common tangent point (Fig. 2.3),
- combined curves—geometric systems consisting of consecutive transition and circular curves (Fig. 2.4).

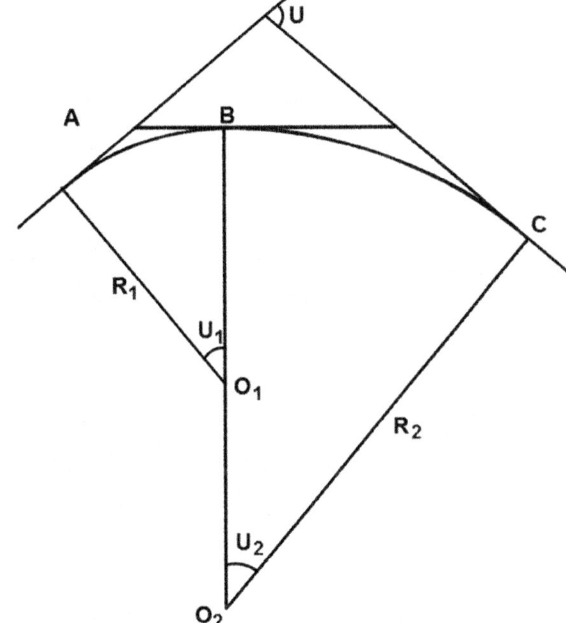

Fig. 2.2 Compound circular curves (*with permission from ASCE*)

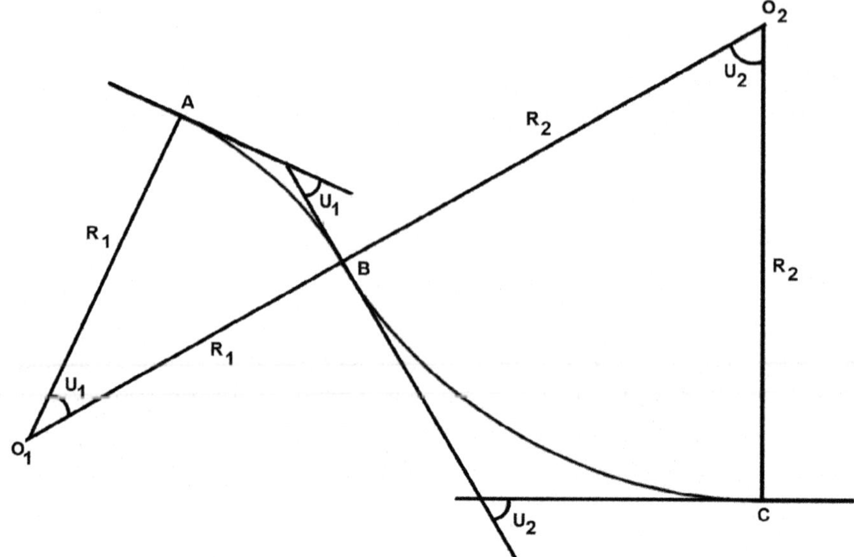

Fig. 2.3 Reverse circular curves (*with permission from ASCE*)

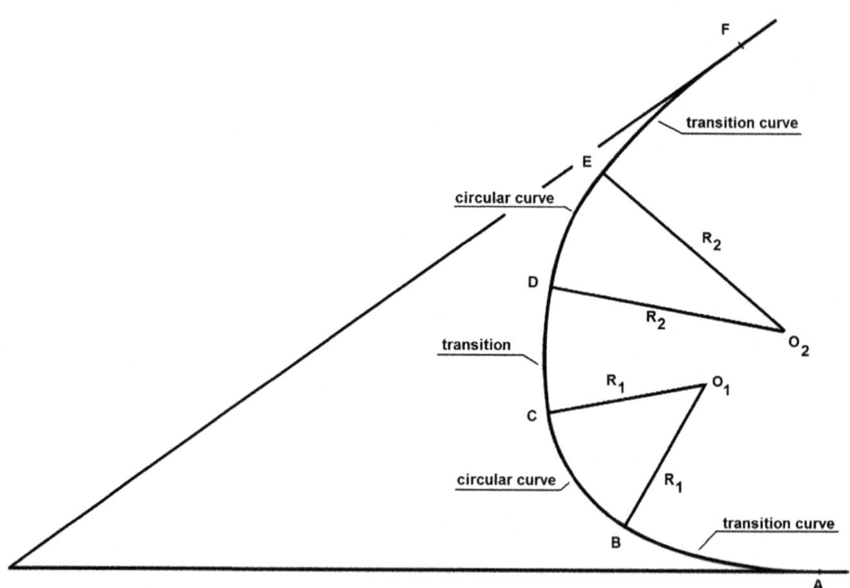

Fig. 2.4 Combined curves (*with permission from ASCE*)

2.2   Parabolic Vertical Curve

Vertical curves are used for to smoothly connection two straight lines with different gradients in the longitudinal profile (Brockenbrough 2009, Easa 2003, Lamm et al. 1999, Meyer and Gibson 1980, Rogers 2008, Wolhuter 2015). In sectional view (Figs. 2.5 and 2.6), the gradient to the left of the vertical curve will be denoted by $p[\%]$ and the gradient to the right will be denoted by $q[\%]$. The vertical curves are generally unsymmetrical and can be crest or sag. It depends on the total change in gradient of two consecutive straight lines:

- crest curves—vertical curves where the total change in gradient is negative,
- sag curves—vertical curves where the total change in gradient is positive.

For some roads (high-speed roads), a cubic parabola is sometimes used as the vertical curve whose rate of change of gradient increases or decreases with the length of the curve. In other cases, a quadratic parabola is generally used as the vertical curve.

In Fig. 2.7 the vertical parabolic curve between two grades p and q which intersect at point W is shown. In this figure are adopted following designations:

P and Q   tangent points,
H the reduced level of P,
L the horizontal length of the curve,
l distance of the highest point of the curve from the point P

The x-y coordinate origin is vertically below P with the x-axis being the datum for reduced levels y.

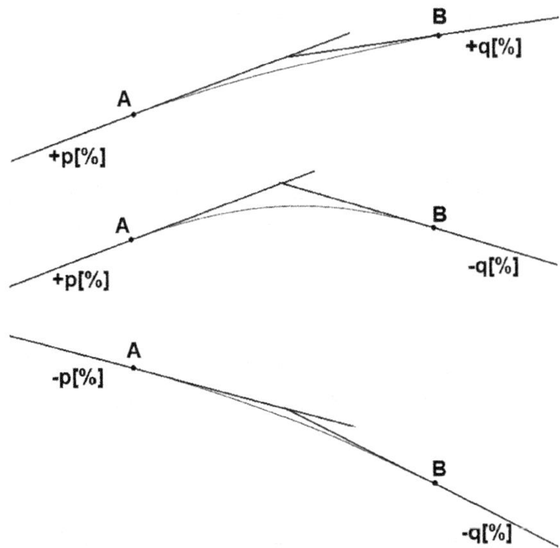

Fig. 2.5   Vertical crest curves

Fig. 2.6 Vertical sag curves

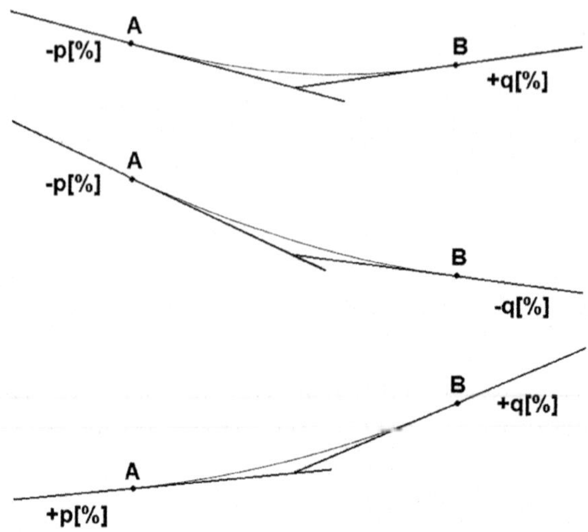

Fig. 2.7 Parabolic vertical curve

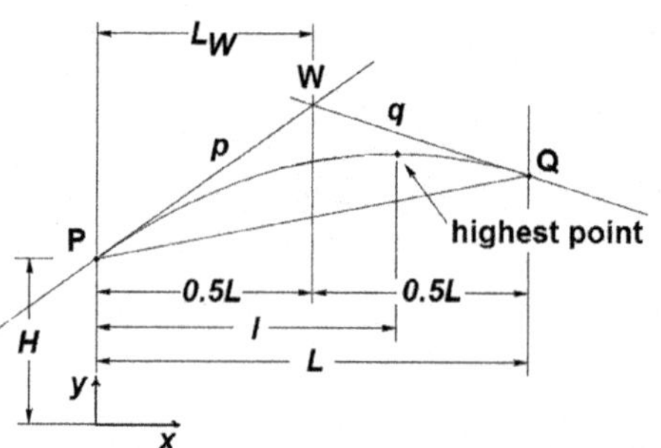

The basic requirement for the vertical curve is that the rate of change of gradient (with respect to horizontal distance) should be constant. The equation of the vertical curve is

$$y = \left(\frac{q - p}{2L}\right)x^2 + px + H \qquad (2.6)$$

The distance of the point W from the point P is

$$L_W = \frac{1}{2}L \qquad (2.7)$$

The horizontal distance to the high point (for crest curve) or low point (for sag curve) is

$$l = \frac{-p}{q-p}L \tag{2.8}$$

The reduced level of point Q is

$$H_q = H + pL_W + q(L - L_W) \tag{2.9}$$

References

Brockenbrough RL (ed) (2009) Highway engineering handbook, 3rd edn. McGraw-Hill, Professional Book Group, New York

Easa SM (2003) Geometric design. In: Chen WF, Liew JYR (eds) The civil engineering handbook. CRC Press, Taylor & Francis Group, Boca Raton

Lamm R, Psarianos B, Mailänder T (1999) Highway design and traffic safety engineering handbook. McGraw-Hill, Professional Book Group, New York

Meyer CF, Gibson DW (1980) Route surveying and design. Harper & Row, New York

Rogers M (2008) Highway engineering, 2nd edn. Wiley-Blackwell, Chichester-Oxford

Wolhuter KM (2015) Geometric design of roads handbook. CRC Press, Taylor & Francis Group, Boca Raton

Chapter 3
Mathematical Methods for Defining of Transition Curves

Transition curves can be defined using the appropriate mathematical formulas that meet a certain boundary conditions. Generally two methods of mathematical description of transition curves are used (Grabowski 1984; Kobryń 2002, 2009):

- using a function of curvature $k = k(l)$ (l—length of the curve),
- using an explicit function $y = f(x)$ in the Cartesian coordinate system.

Another, but very rarely used form of mathematical description of transition curves is to use the function $\rho = \rho(\omega)$ in the polar coordinate system.

3.1 Transition Curves Described Using Curvature Function

Description of transition curves using the function $k = k(l)$ is very convenient to define the specific distribution the curvature within the transition curve. Distribution of the curvature of any transition curve between points P and K always depends on the form of the function $k = k(l)$ that describes the curvature.

Commonly understood transition curves (Fig. 3.1) should allow the gradual introduction of centripetal force on a vehicle moving along the arc. The curvature at the starting point P (for $l = 0$), should be zero:

$$k(l = 0) = 0 \tag{3.1}$$

The curvature within the transition curve should rise to a specified maximum value at the end point K (for $l = L$, where L is the total length of the transition curve):

© Springer International Publishing AG 2017
A. Kobryń, *Transition Curves for Highway Geometric Design*, Springer Tracts on Transportation and Traffic 14, DOI 10.1007/978-3-319-53727-6_3

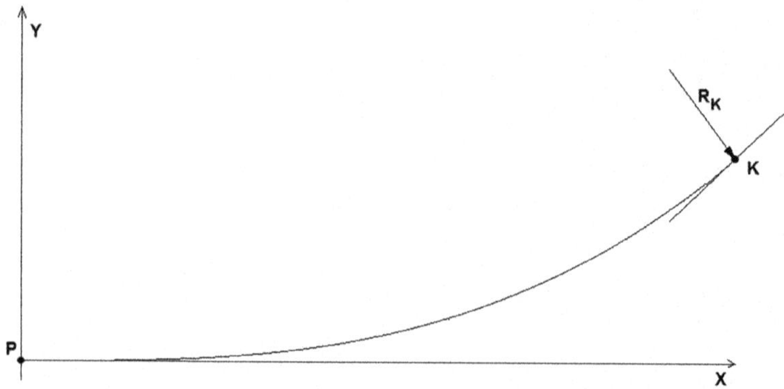

Fig. 3.1 Transition curve

$$k(l = L) = \frac{1}{R_K} \qquad (3.2)$$

wherein R_K is the curvature radius at the point K.

Two physical parameters: normal acceleration and instantaneous change of centrifugal acceleration have an important role when designing the transition curves (Grabowski 1984). The normal acceleration is helpful in determining the required curves radii, so as not to exceed the permissible values of the acceleration. Instantaneous changes of the acceleration along the normal direction to the motion path are helpful in determining the appropriate length of transition curve, which will provide a gentle introduction of centripetal force on a vehicle moving along the arc.

Hence results the need of not exceeding the permissible values of momentary changes of centripetal acceleration. The momentary changes of the acceleration along a normal direction depend on the speed of movement v and of value $\frac{dk}{dl}$. In the case of variable motion speed, they also depend on the value of the acceleration and the curvature of the motion path. Properly designed transition curve should ensure the fulfillment of a condition in the form that not exceed the permissible changes of centripetal acceleration.

In a design practice, this issue are usually considered in a simplified manner, i.e. assuming a constant motion speed. This amounts to determining the permissible value $\frac{dk}{dl}$. The fulfillment of this condition is obtained in practice by setting out of necessary length of the transition curve.

In the case of high speed of motion, transition curve should provide not only the continuity of the curvature, but also the continuity of its changes. This means that at the starting point P and end point K of the transition curve should be:

$$\frac{dk}{dl}(l = 0) = 0 \qquad (3.3)$$

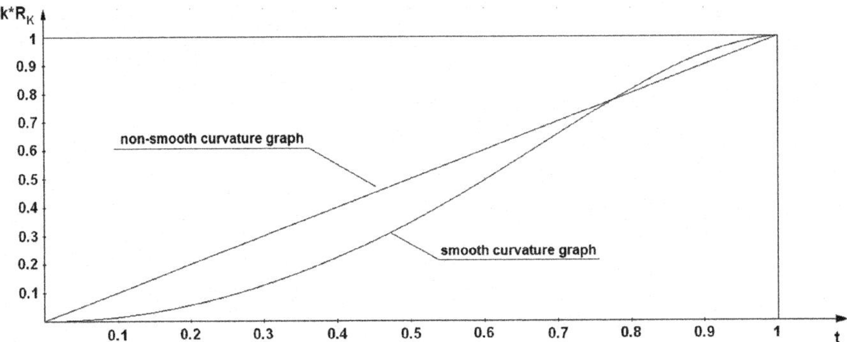

Fig. 3.2 Smooth and noon-smooth curvature graph

$$\frac{dk}{dl}(l = L) = 0 \tag{3.4}$$

The transition curves which fulfill the conditions (3.3) and (3.4), are characterized by the fact that the graph of the curvature does not create any kinks with the adjacent straight and circular arc, which means that the tangent to the graph of the curvature changes its position continuously (Fig. 3.2).

The transition curves that do not only provide continuity of changes in value $\frac{dk}{dl}$, but also continuous changes in value $\frac{d^2k}{dl^2}$, are sometimes contemplated in the literature. This means that at the points P and K is

$$\frac{d^2k}{dl^2}(l = 0) = 0 \tag{3.5}$$

$$\frac{d^2k}{dl^2}(l = L) = 0 \tag{3.6}$$

Solutions of the so-called general transition curves are known from the literature (Grabowski 1984; Kobryń 2002, 2009). They describe a whole curvilinear transition between the two straight lines with only one equation (Fig. 3.3). Thus, they are an alternative to the traditional geometries defined as 1st transition curve—circular curve—2nd transition curve.

The general transition curves fulfill the condition (3.1). The value of the curvature is zero also at the end point:

$$k(l = L) = 0 \tag{3.7}$$

The curvature of the general transition curve reaches the maximum value $1/R_{\min}$ at the point M which is on distance $l = l_M$ from the initial point P:

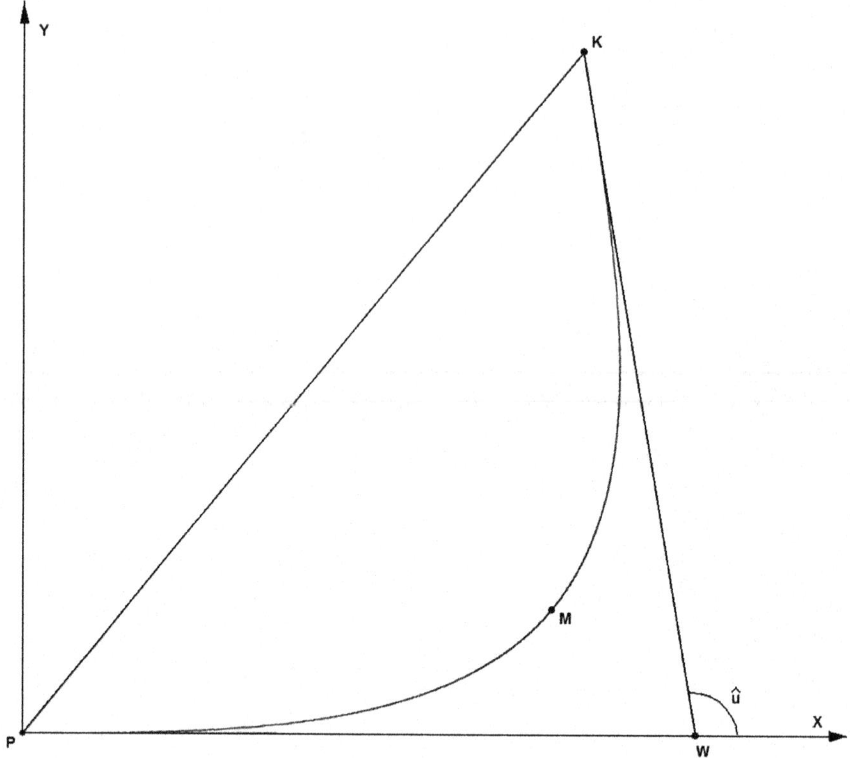

Fig. 3.3 General transition curve

$$k(l = l_M) = 1/R_{\min} \tag{3.8}$$

According to (Grabowski 1984), the general transition curve can also meet the additional conditions, for example (3.3) and (3.4). Depending on the assumptions, the condition $dk/dl = 0$ can be fulfilled additionally at point M:

$$\frac{dk}{dl}(l = l_M) = 0 \tag{3.9}$$

According to (Grabowski 1984), the general transition curve can also fulfill the conditions (2.5) and (2.6).

Ease shaping of curvature is characteristic for the curves defined using the function $k = k(l)$. A shortcoming is the method of determining the coordinates in the Cartesian coordinate system. According to the principles of differential geometry, it is required to use the following formulas to determine the coordinates:

$$x = \int_0^l \cos u \, dl \tag{3.10}$$

$$y = \int_0^l \sin u \, dl \tag{3.11}$$

whereby u is the angle between the tangent at the current point and tangent at the point P. Previously, functions $\cos u$ and $\sin u$ should be expand into power series. The angle u is expressed by formula

$$u = \int_0^l k(l) dl \tag{3.12}$$

Depending on the form of function $k = k(l)$, which determines the curvature of the transition curve, as a rule it leads to a very complex formulas that define the rectangular coordinates. This will be illustrated in Chap. 4.

It should be noted that in practice it is not always necessary to use the full forms of these formulas. Number of members, which are the result of expansion of the function $\cos u$ and $\sin u$ into power series and that should be taken into account when calculating the rectangular coordinates, always depends on the required accuracy of calculation of these coordinates.

3.2   Transition Curves Described Using Explicit Function

These problems associated with the calculation of Cartesian coordinates do not apply to the second way of mathematical description the transition curves, i.e. description using an explicit function $y = f(x)$. In the case of transition curves described using the function $y = f(x)$ a slightly more burdensome analyzes on the distribution of a curvature. The curvature of the curve described by the function $y = f(x)$ can be expressed as:

$$k(x) = \frac{y''}{(1 + y'^2)^{3/2}} \tag{3.13}$$

From Eq. (2.13) it follows that the equivalent conditions to (3.1) and (3.2) for the curves described by the function $y = f(x)$ are following:

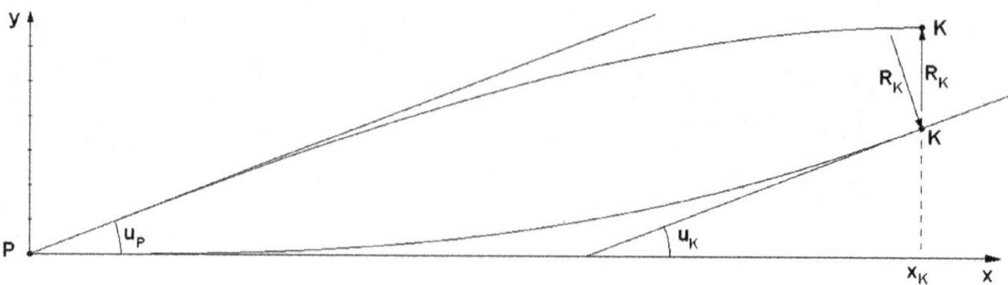

Fig. 3.4 Possible locations of the transition curve in the Cartesian coordinate system

$$y''(x = 0) = 0 \qquad (3.14)$$

$$y''(x = x_K) = \frac{1}{R_K}\left(1 + \tan^2 u_K\right)^{3/2} \qquad (3.15)$$

where:

x_K abscissa of the point K,

$\tan u_K$ tangent inclination at the point K, whereby $\tan u_K = y'(x = x_K)$

Depending on location of the transition curve in the local coordinate system, for the point K are possible two cases (Fig. 3.4):

$$y'(x = x_K) \neq 0 \qquad (3.16)$$

$$y'(x = x_K) = 0 \qquad (3.17)$$

According to Fig. 3.1, a similar options come into play for the point P:

$$y'(x = x_P) \neq 0 \qquad (3.18)$$

$$y'(x = x_P) = 0 \qquad (3.19)$$

In order to define the equivalents of the conditions (3.3) and (3.4) for the curves described using the function $y = f(x)$ should be noted that the value $\frac{dk}{dl}$ may be expressed as:

$$\frac{dk}{dl} = \frac{dk(x)}{dx}\frac{dx}{dl} \qquad (3.20)$$

Taking into account Eq. (3.13), a member $\frac{dk(x)}{dx}$ can be written as:

$$\frac{dk(x)}{dx} = \frac{y''' + y'''y'^2 - 3y'y''^2}{\left(1 + y'^2\right)^{5/2}} \qquad (3.21)$$

According to the principles of differential geometry

$$\frac{dx}{dl} = \frac{1}{(1+y'^2)^{1/2}} \tag{3.22}$$

Therefore, $\frac{dk}{dl}$ finally can be expressed as

$$\frac{dk(x)}{dl} = \frac{y'''(1+y'^2) - 3y'y''^2}{(1+y'^2)^3} \tag{3.23}$$

From Eq. (3.23) it follows that in the case of transition curves described using the function $y = f(x)$, condition (3.3) requires that the counter of expression (3.23) was zero at points P and K of the transition curve. At the point P, a condition

$$\frac{dk(x)}{dl} = 0 \tag{3.24}$$

will be fulfilled if

$$y'''(x = 0) = 0 \tag{3.25}$$

and at the same time the condition (3.17) or condition

$$y''(x = 0) = 0 \tag{3.26}$$

will be fulfilled.

At the point K, the condition (3.24) will be fulfilled if

$$y'''(x = x_K) = 0 \tag{3.27}$$

and at the same time the condition (3.19) or condition

$$y''(x = x_K) = 0 \tag{3.28}$$

will be fulfilled.

3.3   Transition Curves Defined in the Polar Coordinate System

In the case of curves, described using a polar function $\rho = \rho(\omega)$, we are dealing with convenient form from the viewpoint of laying out of these curves (Kobryń 2009). Cartesian coordinates of the curves described by the function $\rho = \rho(\omega)$ can be expressed by the formulas:

$$x = \rho \cos \omega \tag{3.29}$$

$$y = \rho \sin \omega \tag{3.30}$$

In turn, analyzes associated with a distribution curvature and its changes $\frac{dk}{dl}$ require an expression of those values as a function of ρ and ω. According to the principles of differential geometry, the curvature for the curves defined on the basis function $\rho = \rho(\omega)$ is described using the following formula:

$$k(\omega) = \frac{\rho^2 + 2\rho'^2 - \rho\rho''}{(\rho^2 + \rho'^2)^{3/2}} \tag{3.31}$$

Changes in the curvature $\frac{dk(\omega)}{dl}$ can be written as:

$$\frac{dk(\omega)}{dl} = \frac{dk(\omega)}{d\omega}\frac{d\omega}{dl} \tag{3.32}$$

From Eq. (3.31) follows:

$$\frac{dk(\omega)}{d\omega} = \frac{3\rho\rho'\rho''(\rho + \rho'') - \rho'^3(4\rho + 3\rho'') - \rho^3(\rho' + \rho''') - \rho\rho'^2\rho'''}{(\rho^2 + \rho'^2)^{5/2}} \tag{3.33}$$

Since (according to principles of the differential geometry):

$$\frac{d\omega}{dl} = \frac{1}{\sqrt{\rho^2 + \rho'^2}} \tag{3.34}$$

the value $\frac{dk(\omega)}{dl}$ can be written as

$$\frac{dk(\omega)}{dl} = \frac{3\rho\rho'\rho''(\rho + \rho'') - \rho'^3(4\rho + 3\rho'') - \rho^3(\rho' + \rho''') - \rho\rho'^2\rho'''}{(\rho^2 + \rho'^2)^3} \tag{3.35}$$

From the above equations for the curves described by the function $\rho = \rho(\omega)$ results, that the appropriate boundary conditions that define a desired distribution of the curvature $k(\omega)$ and changes $\frac{dk(\omega)}{dl}$ are even more extensive than for the curves $y = f(x)$. With high probability, it can be assumed that this is the reason for the relatively few proposals for solutions of transition curves, which are defined in a such way. Among curves that may have practical significance, only the lemniscate may be taken into account.

References

Grabowski RJ (1984) Gładkie przejścia krzywoliniowe w drogach kołowych i kolejowych. Zeszyty Naukowe AGH, Geodezja nr 82, Kraków (in Polish)

Kobryń A (2002) Wielomianowe krzywe przejściowe w projektowaniu niwelety tras drogowych. Wydawnictwa Politechniki Białostockiej, Rozprawy Naukowe nr 100, Białystok (in Polish)

Kobryń A (2009) Wielomianowe kształtowanie krzywych przejściowych. Wydawnictwa Politechniki Białostockiej, Rozprawy Naukowe nr 167, Białystok (in Polish)

Chapter 4
Transition Curves Described Using Curvature Function

Transition curves defined using the curvature function are easy to shape a curvature. However, a drawback of these curves are very powerful formulas expressing their rectangular coordinates. A spiral curve is a most important curve among the curves which are defined using the curvature function. It is the curve which is commonly used in design of horizontal alignment of roads and highways. It should be noted, however, that other solutions in this group of curves are also known. They will be presented in the following sections of this chapter.

4.1 Classical Transition Curves

4.1.1 Spiral Curve

Clothoid (also known as Cornu spiral or a spiral curve) is described by so-called natural equation (Lamm et al. 1999; Lipiński 1993; Lorenz 1971, Meyer and Gibson 1980), which has following form:

$$a^2 = r \cdot l = const. \tag{4.1}$$

where:

a so-called parameter of the spiral curve,
r radius of curvature r at any clothoid point,
l arc length measured from the initial point (natural parameter)

Graph of the curve defined by Eq. (4.1) is shown in Fig. 4.1. Only the first part of the curve in the positive range of coordinates x and y is used in the design practice.

Apart from the curvature radius r and the curve length l, an important parameter that describes the geometry of the spiral curve is a deflection angle, which will be

© Springer International Publishing AG 2017

A. Kobryń, *Transition Curves for Highway Geometric Design*, Springer Tracts on Transportation and Traffic 14, DOI 10.1007/978-3-319-53727-6_4

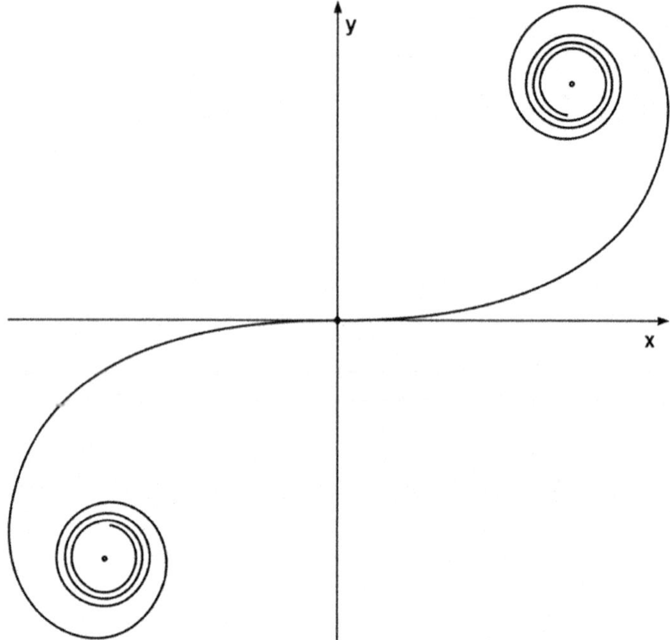

Fig. 4.1 Spiral curve

denoted as u. As it is known, the value of this angle is given by following formulas (e.g. Lamm et al. 1999; Lipiński 1993; Lorenz 1971, Meyer and Gibson 1980):

$$u = \frac{l^2}{2a^2} = \frac{l}{2r} = \frac{a^2}{2r^2} \tag{4.2}$$

Formulas (4.2) describe an interrelationships between parameter of the clothoid (a), its length (l), the curvature radius (r) and the deflection angle (u).

Cartesian coordinates of the spiral curve can be expressed using known formulas:

$$x = l\left(1 - \frac{l^4}{40a^4} + \frac{l^8}{3456a^8} - \frac{l^{12}}{599040a^{12}} + \cdots\right) \tag{4.3}$$

$$y = l\left(\frac{l^2}{6a^2} - \frac{l^6}{336a^6} + \frac{l^{10}}{42240a^{10}} - \frac{l^{14}}{9676800a^{14}} + \cdots\right) \tag{4.4}$$

or

$$x = l\left[1 - \frac{1}{10}\left(\frac{l}{2r}\right)^2 + \frac{1}{216}\left(\frac{l}{2r}\right)^4 - \frac{1}{9360}\left(\frac{l}{2r}\right)^6 + \cdots\right] \tag{4.5}$$

$$y = l\left[\frac{1}{3}\left(\frac{l}{2r}\right) - \frac{1}{42}\left(\frac{l}{2r}\right)^3 + \frac{1}{1320}\left(\frac{l}{2r}\right)^5 - \frac{1}{75600}\left(\frac{l}{2r}\right)^7 + \cdots\right] \tag{4.6}$$

Taking into account the formulas (4.2), according to which is $u = \frac{l}{2r}$, the Eqs. (4.3) and (4.4) can be expressed as:

$$x = l\left(1 - \frac{u^2}{10} + \frac{u^4}{216} - \frac{u^6}{9360} + \cdots\right) \tag{4.7}$$

$$y = l\left(\frac{u}{3} - \frac{u^3}{42} + \frac{u^5}{1320} - \frac{u^7}{75600} + \cdots\right) \tag{4.8}$$

According to needs, in the engineering practice is generally sufficient restriction to two or three members, which occur in the above formulas. This results from the small deflection angles ($u \leq 0.1$ rad) in the geometrical systems which are used in the applications of the clothoid In highway design.

In the work (Grabowski and Kobryń 1997) an alternative form of the clothoid equation were proposed:

$$k(l) = \frac{1}{R_K}\frac{l}{L} \tag{4.9}$$

where R_K—minimal radius of curvature, L—total length, l—current length of the spiral curve (natural parameter).

On the basis of Eq. (3.10), the deflection angle at any point located at the distance l from the start point can be expressed as:

$$u = \frac{1}{2}\frac{L}{R_K}\left(\frac{l}{L}\right)^2 \tag{4.10}$$

After adoption of a designation $t = l/L$, Eq. (4.10) takes a following form:

$$u = \frac{1}{2}\frac{L}{R_K}t^2 \tag{4.11}$$

From the above formula it follows that the deflection angle u at the end point of the spiral curve, i.e. for $l = L$, is:

$$\hat{u} = \frac{L}{2R_K} \tag{4.12}$$

The formula (4.12) plays an important role in determining the relationships between the geometrical parameters of the spiral curve (Grabowski and Kobryń 1997).

According to (Grabowski and Kobryń 1997), the Cartesian coordinates for any point of the spiral curve can be expressed by means of following formulas:

$$x = L\left[t - \frac{1}{40} s^2 t^5 + \frac{1}{3456} s^4 t^9 - \cdots \right] \tag{4.13}$$

$$y = L\left[\frac{1}{6} s t^3 - \frac{1}{336} s^3 t^7 + \frac{1}{42240} s^5 t^{11} - \cdots \right] \tag{4.14}$$

whereby: $t = l/L$, $s = L/R_K$.

Substituting $s = L/R_K$ and taking into account Eq. (4.12), the formulas (4.13) and (4.14) can be written appropriately as:

$$x = L\left[t - \frac{1}{10}\left(\frac{L}{2R_K}\right)^2 t^5 + \frac{1}{216}\left(\frac{L}{2R_K}\right)^4 t^9 - \cdots \right] \tag{4.15}$$

$$y = L\left[\frac{1}{3}\left(\frac{L}{2R_K}\right) t^3 - \frac{1}{42}\left(\frac{L}{2R_K}\right)^3 t^7 - + \frac{1}{1320}\left(\frac{L}{2R_K}\right)^5 t^{11} - \cdots \right] \tag{4.16}$$

or

$$x = L\left[t - \frac{1}{10}\hat{u}^2 t^5 + \frac{1}{216}\hat{u}^4 t^9 - \cdots \right] \tag{4.17}$$

$$y = L\left[\frac{1}{3}\hat{u} t^3 - \frac{1}{42}\hat{u}^3 t^7 + \frac{1}{1320}\hat{u}^5 t^9 - \cdots \right] \tag{4.18}$$

The Cartesian coordinates X and Y at the end point of the clothoid (i.e. for $t = 1$) can be expressed on the basis of (4.15) and (4.16) as follows:

$$X = L\left[1 - \frac{1}{10}\left(\frac{L}{2R_K}\right)^2 + \frac{1}{216}\left(\frac{L}{2R_K}\right)^4 - \cdots \right] \tag{4.19}$$

$$Y = L\left[\frac{1}{3}\left(\frac{L}{2R_K}\right) - \frac{1}{42}\left(\frac{L}{2R_K}\right)^3 + \frac{1}{1320}\left(\frac{L}{2R_K}\right)^5 - \cdots \right] \tag{4.20}$$

Whereas, from formulas (4.17) and (4.18) are obtained:

$$X = L\left[1 - \frac{1}{10}\hat{u}^2 + \frac{1}{216}\hat{u}^4 - \cdots\right] \qquad (4.21)$$

$$Y = L\left[\frac{1}{3}\hat{u} - \frac{1}{42}\hat{u}^3 + \frac{1}{1320}\hat{u}^5 - \cdots\right] \qquad (4.22)$$

As can be seen, these four formulas are an equivalents of the formulas (4.3) and (4.4), as well as (4.7) and (4.8).

Formulas (4.13)–(4.18) are very useful for determining the coordinates of a greater number of points within the spiral curve. With their help it is possible to easily define the position of any point relative to the beginning or end the spiral curve, allowing you to automate a calculation process. It should be noted also that—with the assumed value of the ratio $s = L/R_K$ (or the angle $\hat{u}$)—the values occurring in square brackets for each values of $t \in \,<0; 1>$ are constant, regardless of the length of the transition curve. When $L = 1$, a particular formulas refer to the curve which is called in a specialist literature as the unit clothoid.

4.1.2 **Bloss** *Curve*

The clothoid fulfills the basic design conditions of the transition curve, i.e. (3.1) and (3.2). The curve, which in addition to these conditions also satisfies the conditions (3.3) and (3.4), is so-called *Bloss* curve. The equations necessary to determine the curve are given in the work (Grabowski and Kobryń 1997).

An equation of the *Bloss* curve has a following form:

$$k(l) = \frac{1}{R_K}\left[3\left(\frac{l}{L}\right)^2 - 2\left(\frac{l}{L}\right)^3\right] \qquad (4.23)$$

A deflection angle for the curve (4.23) is described by an equation:

$$u = \frac{L}{R_K}\left[\left(\frac{l}{L}\right)^3 - \frac{1}{2}\left(\frac{l}{L}\right)^4\right] \qquad (4.24)$$

From the above equation results that the angle u at the end point, i.e. for $l = L$, is equal to $\hat{u} = \frac{L}{2R_K}$ (as in the case of the spiral curve).

Using (3.8), (3.9) and (4.24), expanding the functions $\cos u$ and $\sin u$ in a power series and integrating, the formulas describing the Cartesian coordinates of the curve (4.23) are obtained. Appropriate formulas were presented, among others, in the papers (Grabowski and Kobryń 1997; Kobryń 1991a, 2008). They are as follows:

$$x = L\left[t - s^2 P_2 t^3 + s^4 P_4 t^5 - s^6 P_6 t^7 + \cdots\right] \tag{4.25}$$

$$y = L\left[s P_1 t^2 - s^3 P_3 t^4 + s^5 P_5 t^6 - s^7 P_7 t^8 + \cdots\right] \tag{4.26}$$

where

$$P_1 = \frac{1}{4} t^2 - \frac{1}{10} t^3$$

$$P_2 = \frac{1}{14} t^4 - \frac{1}{16} t^5 + \frac{1}{72} t^6$$

$$P_3 = \frac{1}{60} t^6 - \frac{1}{44} t^7 + \frac{1}{96} t^8 - \frac{1}{624} t^9$$

$$P_4 = \frac{1}{312} t^8 - \frac{1}{168} t^9 + \frac{1}{240} t^{10} - \frac{1}{768} t^{11} + \frac{1}{6528} t^{12}$$

$$P_5 = \frac{1}{1920} t^{10} - \frac{1}{816} t^{11} + \frac{1}{864} t^{12} - \frac{1}{1824} t^{13} + \frac{1}{7680} t^{14} - \frac{1}{80640} t^{15}$$

$$P_6 = \frac{1}{13680} t^{12} - \frac{1}{4800} t^{13} + \frac{1}{4032} t^{14} - \frac{1}{6336} t^{15} + \frac{1}{17664} t^{16}$$
$$- \frac{1}{92160} t^{17} + \frac{1}{1152000} t^{18}$$

$$P_7 = \frac{1}{110880} t^{14} - \frac{1}{33120} t^{15} + \frac{1}{23040} t^{16} - \frac{1}{28800} t^{17} + \frac{1}{59904} t^{18}$$
$$- \frac{1}{207360} t^{19} + + \frac{1}{1290240} t^{20} - \frac{1}{187081480} t^{21} + \frac{1}{1290240} t^{20}$$
$$- \frac{1}{187081480} t^{21}$$

whereby $t = l/L$ ($l \leq L$) and $s = L/R_K$.

As can be seen, formulas describing the rectangular coordinates of the *Bloss* curve are relatively more complex than in the case of the spiral curve. However, according to (Grabowski and Kobryń 1997), it should be noted that it is generally possible to skip the members containing P_6 and P_7 (without a prejudice to the accuracy of the coordinates calculation).

4.1.3 **Grabowski** *Curve*

The curve, which in addition to the conditions fulfilled by the *Bloss* curve, also meets the conditions (3.5) and (3.6) is presented in the paper (Grabowski 1973). Its equation is as follows:

$$k(l) = \frac{1}{R_K}\left[10\left(\frac{l}{L}\right)^3 - 15\left(\frac{l}{L}\right)^4 + 6\left(\frac{l}{L}\right)^5\right] \tag{4.27}$$

The deflection angle is expressed as

$$u = \frac{L}{R_K}\left[\frac{5}{2}\left(\frac{l}{L}\right)^4 - 3\left(\frac{l}{L}\right)^5 + \left(\frac{l}{L}\right)^6\right] \tag{4.28}$$

Similarly as for the clothoid and the *Bloss* curve, the angle u at the end point of the curve (i.e. for $l = L$) is equal to $\hat{u} = \frac{L}{2R_K}$.

According to (Kobryń 2008), formulas describing rectangular coordinates of the curve (4.27) are as follows:

$$x = L\left[t - s^2 Q_2 t^3 + s^4 Q_4 t^5 - s^6 Q_6 t^7 + \cdots\right] \tag{4.29}$$

$$y = L\left[s Q_1 t^2 - s^3 Q_3 t^4 + s^5 Q_5 t^6 - s^7 Q_7 t^8 + \cdots\right] \tag{4.30}$$

where:

$$Q_1 = \frac{1}{2}t^3 - \frac{1}{2}t^4 + \frac{1}{7}t^5$$

$$Q_2 = \frac{25}{72}t^6 - \frac{3}{4}t^7 + \frac{7}{11}t^8 - \frac{1}{4}t^9 + \frac{1}{26}t^{10}$$

$$Q_3 = \frac{125}{624}t^9 - \frac{75}{112}t^{10} + \frac{23}{24}t^{11} - \frac{3}{4}t^{12} + \frac{23}{68}t^{13} - \frac{1}{12}t^{14} + \frac{1}{114}t^{15}$$

$$Q_4 = \frac{625}{6528}t^{12} - \frac{125}{288}t^{13} + \frac{50}{57}t^{14} - \frac{33}{32}t^{15} + \frac{37}{48}t^{16} - \frac{3}{8}t^{17} + \frac{8}{69}t^{18} - \frac{1}{48}t^{19} + \frac{1}{600}t^{20}$$

$$Q_5 = \frac{625}{16128}t^{15} - \frac{625}{2816}t^{16} + \frac{5125}{8832}t^{17} - \frac{175}{192}t^{18} + \frac{457}{480}t^{19} - \frac{1437}{2080}t^{20} + \frac{457}{1296}t^{21} - \frac{1}{8}t^{22} + \frac{41}{1392}t^{23} - \frac{1}{240}t^{24} + \frac{1}{3720}t^{25}$$

$$Q_6 = \frac{125}{9216}t^{18} - \frac{625}{6656}t^{19} + \frac{3125}{10368}t^{20} - \frac{2125}{3584}t^{21} + \frac{17725}{22272}t^{22} - \frac{737}{960}t^{23}$$
$$+ \frac{12179}{22320}t^{24} - \frac{737}{2560}t^{25} + \frac{709}{6336}t^{26} - \frac{1}{32}t^{27} + \frac{1}{168}t^{28} - \frac{1}{1440}t^{29}$$
$$+ \frac{1}{79920}t^{30}$$

$$Q_7 = \frac{15625}{3741696}t^{21} - \frac{4375}{129024}t^{22} + \frac{36875}{285696}t^{23} - \frac{625}{2048}t^{24} + \frac{25375}{50688}t^{25}$$
$$- \frac{5245}{8704}t^{26} + \frac{44257}{80640}t^{27} - \frac{11617}{30240}t^{28} + \frac{44257}{213120}t^{29} - \frac{7343}{85120}t^{30}$$
$$+ \frac{203}{7488}t^{31} - \frac{1}{160}t^{32} + \frac{59}{59040}t^{33}$$

wherein in the all above equations earlier notations: $t = l/L$, $s = L/R_K$ are used.

4.1.4   Other Transition Curves Described Using Curvature Function

In addition to the curves (4.23) and (4.27), other curves described function of the curvature are also known from the literature. These are:

- *Auberlen* curve (Auberlen 1956)

$$k(l) = \frac{1}{2R_K}\left[1 - \cos\left(\Pi\frac{l}{L}\right)\right] \qquad (4.31)$$

- *Klein* curve (Klein 1937)

$$k(l) = \frac{1}{R_K}\left[\frac{l}{L} - \frac{1}{2\Pi}\sin\left(2\Pi\frac{l}{L}\right)\right] \qquad (4.32)$$

According to (Kobryń 2008), the deflection angle is expressed by following formulas:

- for the curve (4.31)

$$u = \frac{1}{2}\frac{L}{R_K}\left[\frac{l}{L} - \frac{1}{\Pi}\sin\left(\Pi\frac{l}{L}\right)\right] \qquad (4.33)$$

- for the curve (4.32)

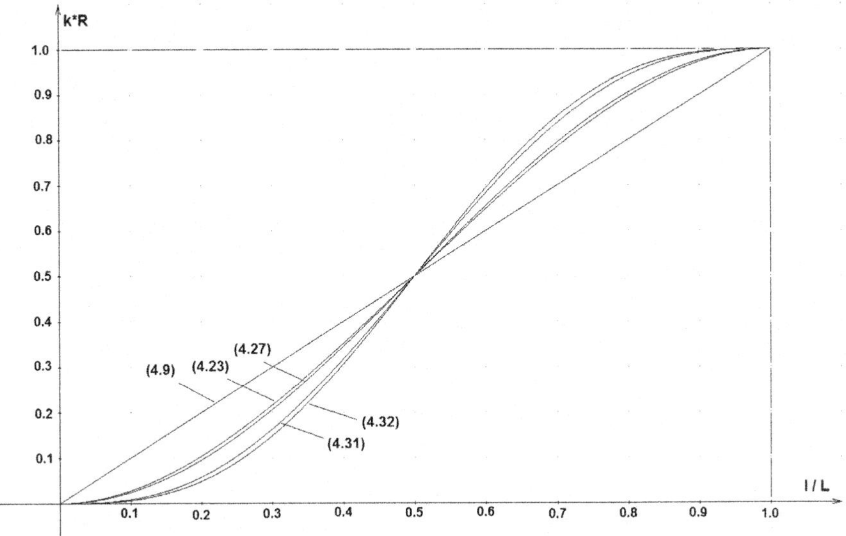

Fig. 4.2 Graphs of curvature for selected curves described by the curvature function

$$u = \frac{1}{2}\frac{L}{R_K}\left[\left(\frac{l}{L}\right)^2 + \frac{1}{2\Pi^2}\left(\cos\left(2\Pi\frac{l}{L}\right) - 1\right)\right] \tag{4.34}$$

While the curve (4.31) fulfills the conditions (3.3) and (3.4) at the points P and K ($\frac{dk}{dl} = 0$), at points P and K of the curve (4.33) is additionally $\frac{d^2k}{dl^2} = 0$.

It should be noted that the formulas describing the rectangular coordinates for curves (4.31) and (4.33) are even more complex than for the appropriate curves (4.23) and (4.27).

It should also be noted that the distribution of the curvature as well as a distribution of $\frac{dk}{dl}$ for the curve (4.31) are similar to the curve (4.23), while for the curve (4.32)—to the curve (4.27). This is shown in Fig. 4.2. Additionally, Fig. 4.3 shows graphs of $\frac{dk}{dl}$, which are described by following formulas:

- for the spiral curve (4.9)

$$\frac{dk}{dl} = \frac{1}{R_K L} = \text{const.} \tag{4.35}$$

- for the *Bloss* curve (4.23)

$$\frac{dk}{dl} = \frac{6}{RL}\left(t - t^2\right) \tag{4.36}$$

- for the *Grabowski* curve (4.27)

$$\frac{\mathrm{d}k}{\mathrm{d}l} = \frac{30}{R_K L}\left(t^2 - 2t^3 + t^4\right)$$
(4.37)

- for the *Auberlen* curve (4.31)

$$\frac{\mathrm{d}k}{\mathrm{d}l} = \frac{\Pi}{2R_K L}\sin(\Pi t)$$
(4.38)

- for the *Klein* curve (4.32)

$$\frac{\mathrm{d}k}{\mathrm{d}l} = \frac{1}{R_K L}\left(1 - \cos(2\Pi t)\right)$$
(4.39)

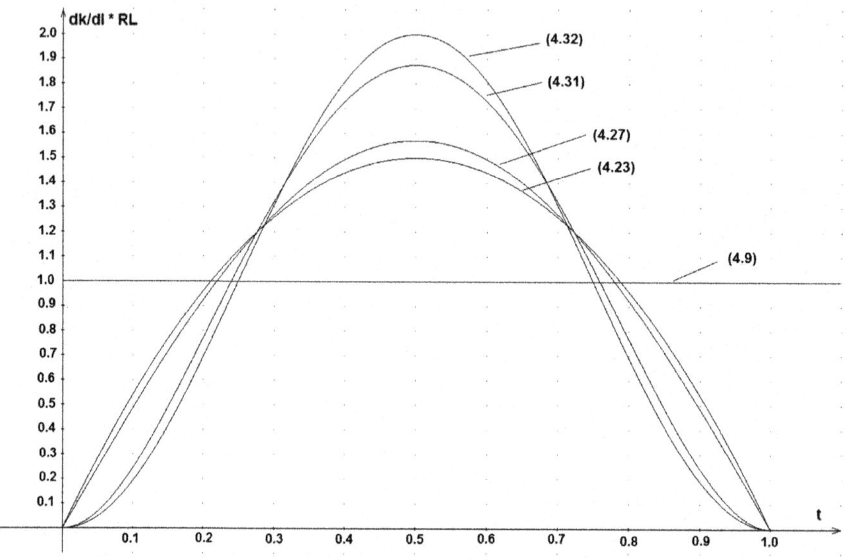

Fig. 4.3 Graphs of values dk/dl for selected curves described by the curvature function

Table 4.1 Values dk/dl at selected points of curves described by the curvature function $k = k(l)$

Curve	$l = 0$	$l = \frac{1}{4}L$	$l = \frac{1}{2}L$	$l = \frac{3}{4}L$	$l = L$
Curve (4.9)	$k = 0$	$k = \frac{1}{4}\frac{1}{R_K}$	$k = \frac{1}{2}\frac{1}{R_K}$	$k = \frac{3}{4}\frac{1}{R_K}$	$k = \frac{1}{R_K}$
Curve (4.23)	$k = 0$	$k = \frac{5}{32}\frac{1}{R_K}$	$k = \frac{1}{2}\frac{1}{R_K}$	$k = \frac{27}{32}\frac{1}{R_K}$	$k = \frac{1}{R_K}$
Curve (4.27)	$k = 0$	$k = \frac{2-\sqrt{2}}{4}\frac{1}{R_K}$	$k = \frac{1}{2}\frac{1}{R_K}$	$k = \frac{2+\sqrt{2}}{4}\frac{1}{R_K}$	$k = \frac{1}{R_K}$
Curve (4.31)	$k = 0$	$k = \frac{53}{512}\frac{1}{R_K}$	$k = \frac{1}{2}\frac{1}{R_K}$	$k = \frac{459}{512}\frac{1}{R_K}$	$k = \frac{1}{R_K}$
Curve (4.32)	$k = 0$	$k = \left(\frac{1}{4}-\frac{1}{2\Pi}\right)\frac{1}{R_K}$	$k = \frac{1}{2}\frac{1}{R_K}$	$k = \left(\frac{3}{4}+\frac{1}{2\Pi}\right)\frac{1}{R_K}$	$k = \frac{1}{R_K}$

Based on the graphs it can be seen that the curve (4.31) bears a strong resemblance to the *Bloss* curve (4.23), while the curve (4.32)—to the *Grabowski* curve (4.27). At the same time, with an increase in degree of smoothness of the curvature graph, a milder changes of the curvature at vicinity of the start point and the end point are shown. However, significantly greater curvature changes at centre of the curve arc are a consequence of this.

Table 4.1 shows the values of dk/dl at selected points in each curve. On the basis of the maximum values of dk/dl, which are shown in gray in Table 4.1, a required minimum length of each transition curve can be determined. According to (Kobryń 2008), they achieve values given in Table 4.2.

A detailed analysis confirms an earlier signaled similarity of appropriate curves. A difference of right sides of Eqs. (4.23) and (4.31) is

$$\Delta k = \frac{1}{R_K}\left[3\left(\frac{l}{L}\right)^2 - 2\left(\frac{l}{L}\right)^3\right] - \frac{1}{2R_K}\left[1 - \cos\left(\Pi\frac{l}{L}\right)\right] \tag{4.40}$$

It follows from the necessary condition (i.e. $(\Delta k)' = 0$) that the value ΔK is maximal if

$$12t - 12t^2 - \Pi\,\sin(\Pi t) = 0 \tag{4.41}$$

Because a graph of the curvature is symmetrical with respect to the point described by the parameter $l = L/2$ (Table 4.1), Eq. (4.41) is fulfilled in two cases (i.e. for $t_1 = 0.278$ and $t_2 = 0.722$). A maximum value of the expression (4.40) is $\Delta k = 0.010\frac{1}{R_K}$.

A similar analysis can be made for curves (4.27) and (4.32). A difference of right sides of these equations is:

Table 4.2 The minimum length of transition curves resulting from limit values for changes of the centrifugal acceleration

Curve	Minimal length of the curve	Elongation of the curve relative to the clothoid
Curve (4.9)	$L = \dfrac{v^3}{R_K p_{max}}$	–
Curve (4.23)	$L = \dfrac{3}{2}\dfrac{v^3}{R_K p_{max}}$	1.500
Curve (4.27)	$L = \dfrac{1}{2}\Pi\dfrac{v^3}{R_K p_{max}}$	1.571
Curve (4.31)	$L = \dfrac{30}{16}\dfrac{v^3}{R_K p_{max}}$	1.875
Curve (4.32)	$L = 2\dfrac{v^3}{R_K p_{max}}$	2.000

$$\Delta k = \left[10\left(\frac{l}{L}\right)^3 - 15\left(\frac{l}{L}\right)^4 + 6\left(\frac{l}{L}\right)^5 - \frac{l}{L} + \frac{1}{2\Pi}\sin\left(2\Pi\frac{l}{L}\right) \right] \tag{4.42}$$

On the basis of the condition $(\Delta k)' = 0$ is a following equation obtained:

$$30t^2 - 60t^3 + 30t^4 + \cos(2\Pi t) - 1 = 0 \tag{4.43}$$

Eq. (4.43) has two solutions (for $t_1 = 0.315$ and $t_2 = 0.685$) corresponding to maximum value of $\Delta k = 0.015\frac{1}{R_K}$.

In both cases, therefore, the differences Δk are insignificant. For this reason, it will be preferred a practical usefulness of these curves, which have a simpler form of mathematical formulas expressing the Cartesian coordinates. Disregarding the clothoid, that is definitely the most advantageous in this regard, better satisfy this criterion curves (4.23) and (4.27) (vs. (4.31) and (4.32)).

4.1.5 Two-Parameter Spiral Curves

Using a similar notation as for the classic spiral curve, an equation of two-parameter clothoids can be written as (Lorenz 1971; Lipiński 1993):

$$a^{n+1} = r \cdot l^n \tag{4.44}$$

Apart from the number a, the exponent n is the second parameter that changes a curve geometry. Two-parameter spiral curves have been proposed to design of special sections of road interchanges where there are small values of curvature radii and large changes of motion speed.

The paper (Kobryń 2008) presents an alternative form of equation for two-parameter clothoids:

$$k(l) = \frac{1}{R_K}\left(\frac{l}{L}\right)^n \tag{4.45}$$

Eq. (4.45) for $n = 1$ takes the form (4.9) and describes the classical clothoid. Graphs of curvature for the curves described by Eq. (4.45) for selected values of n are show in Fig. 4.4. Whereas Fig. 4.5 shows graphs of values dk/dl, which are expressed as

$$\frac{dk}{dl} = n\frac{1}{R_K L}t^{n-1} \tag{4.46}$$

The deflection angle in the case of the curves (4.45) is described by following equation:

$$u = \frac{1}{n+1}\frac{L}{R_K}\left(\frac{l}{L}\right)^{n+1} \tag{4.47}$$

The deflection angle at the end point (for $l = L$) is:

$$\hat{u} = \frac{1}{n+1}\frac{L}{R_K} \tag{4.48}$$

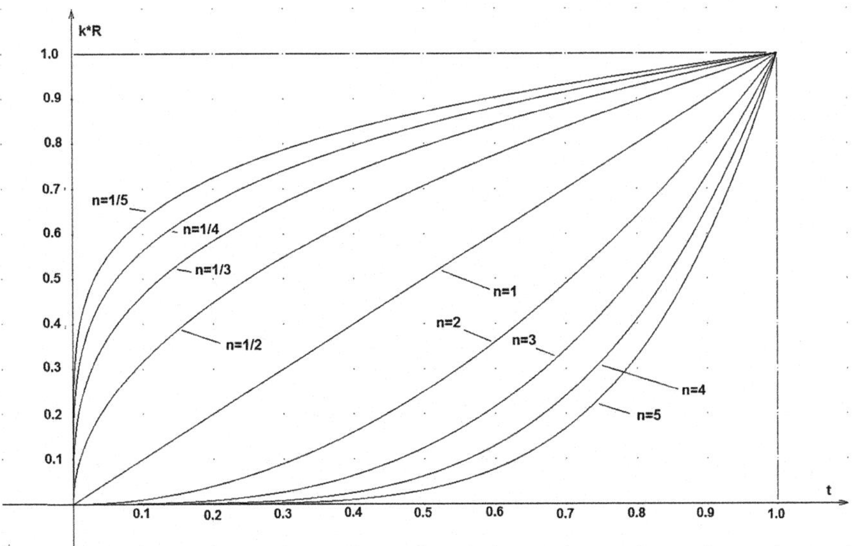

Fig. 4.4 Graphs of curvature for curves (4.45)

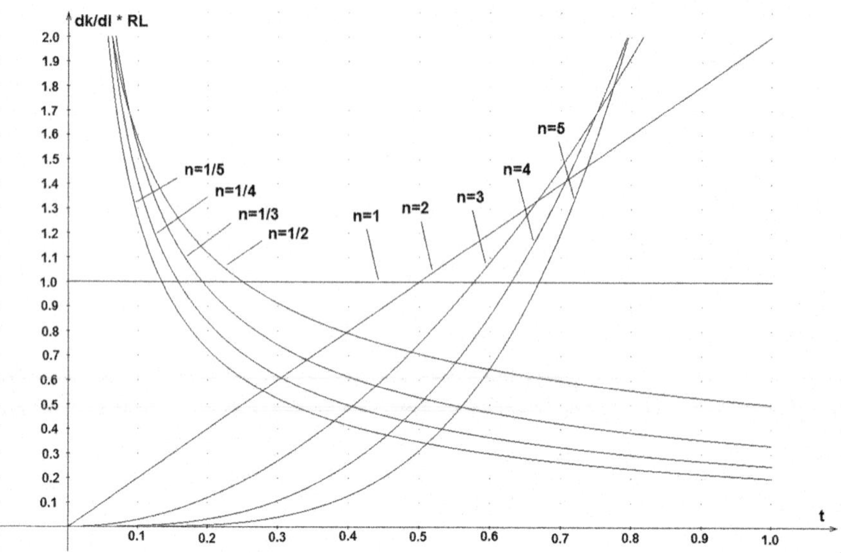

Fig. 4.5 Graphs of values dk/dl for curves (4.45)

The rectangular coordinates of the curves (4.45) can be calculated using following formulas (Kobryń 1991b, 1995, 2008):

$$x = L\left[t - \frac{1}{2!(n+1)^2(2n+3)}s^2 t^{2n+3} + \frac{1}{4!(n+1)^4(4n+5)}s^4 t^{4n+5} \right.$$
$$\left. - \frac{1}{6!(n+1)^6(6n+7)}s^6 t^{6n+7} + \cdots\right] \tag{4.49}$$

$$y = L\left[\frac{1}{1!(n+1)(n+2)}s t^{n+2} - \frac{1}{3!(n+1)^3(3n+4)}s^3 t^{3n+4} \right.$$
$$\left. + \frac{1}{5!(n+1)^5(5n+6)}s^5 t^{5n+6} - \frac{1}{7!(n+1)^7(7n+8)}s^7 t^{7n+8} + \cdots\right] \tag{4.50}$$

whereby: $t = l/L$, $s = L/R_K$.

4.2 Vertical Transition Curve Described Using Curvature Function

In the works (Kobryń 1999, 2007) a solution of the transition curve dedicated to design of vertical curves are presented. This curve is described by a function of curvature and was designated taking into account the terms of the motion dynamics. In accordance with those conditions, an acceleration vector $\vec{a}$ (located in the osculating plane to the curve which is a motion path) is described by following equation (Leyko 1996):

$$\vec{a} = \frac{dv}{dt}\vec{t} + v^2 k\vec{n} \tag{4.51}$$

where: v—motion speed, k—curvature of the trajectory, $\vec{t}$ and $\vec{n}$—appropriately are tangent versor and unit normal of the Frenet trihedron.

A derivative of the function (4.51) is given by:

$$\vec{p} = \left(\frac{d^2 v}{dt^2} - v^3 k^2\right)\vec{t} + \left(3v\frac{dv}{dt}k + v^3\frac{dk}{dl}\right)\vec{n} + v^3 Tk\vec{b} \tag{4.52}$$

where: l—natural parameter of trajectory, T—trajectory torsion (deflection), $\vec{b}$—binormal versor of the Frenet trihedron,

Taking into account the movement within the vertical curve and considering the effect of gravity, vector of momentary changes of the centripetal acceleration can be written as follows (Kobryń 1999):

$$\vec{p} = \left(\frac{d^2 v}{dt^2} - v^3 k^2 + vgk\,\cos\beta\right)\vec{t} + \left(3v\frac{dv}{dt}k + v^3\frac{dk}{dl} - vgk\,\sin\beta\right)\vec{n} \\ + (v^3 Tk - vgT\,\cos\beta)\vec{b} \tag{4.53}$$

A component of this vector in normal direction is described by following equation:

$$p_n = 3v\frac{dv}{dt}k + v^3\frac{dk}{dl} - vgk\,\sin\beta \tag{4.54}$$

Assuming a permissible value of p_n on the vertical arc, Eq. (4.54) can be written as:

$$3v\frac{dv}{dt}k + v^3\frac{dk}{dl} - vgk\,\sin\beta = Q \tag{4.55}$$

Let:

$$A = 3va - vg \sin \beta$$

$$B = v^3$$

wherein: $a = \frac{\mathrm{d}v}{\mathrm{d}t}$ (acceleration).

After appropriate transformations, Eq. (4.55) takes a following form:

$$\frac{\mathrm{d}k}{\mathrm{d}l} = \frac{Q}{B} - \frac{A}{B}k \tag{4.56}$$

After solving this differential equation, the following equation is obtained (Kobryń 1999):

$$k(l) = \frac{Q}{A}\left(1 - e^{-\frac{A}{B}l}\right) \tag{4.57}$$

The above equation describes the vertical transition curve.

Assuming simplified assumption that the speed is constant ($a = 0$), and excluding the impact of gravity, Eq. (4.57) can be written as:

$$v^3 \frac{\mathrm{d}k}{\mathrm{d}l} = Q \tag{4.58}$$

It follows:

$$k(l) = \frac{Q}{v^3}l \tag{4.59}$$

As can be seen, the above equation describes a commonly known clothoid. In this case, its parameter is equal to value v^3/Q.

Eqs. (4.57) and (4.59) determine the curvature of the vertical transition curves (according to the assumptions). From these equations follows an appropriate minimum length of the vertical transition curve:

$$L = \frac{B}{A} \ln \frac{R_K Q}{R_K Q - A} \tag{4.60}$$

or

$$L = \frac{v^3}{Q} \frac{1}{R_K} \tag{4.61}$$

So far, a problem of permissible changes of centripetal acceleration on vertical curves is not fully solved. This problem was discussed in a literature rarely. Only few works give a specific numerical value, for example: $Q = 0.24 \, \text{m/s}^3$ (Durth 1974), $Q = 0.3 \, \text{m/s}^3$ (McConell 1957), $Q = 0.25 \, \text{m/s}^3$ (Melchior 1928).

It turns out that the differences in the required length of the vertical transition curves calculated from Eqs. (4.60) and (4.61) with $Q = 0.25 \, \text{m/s}^3$ are very small (Table 4.3).

This allows during further considerations to limit using of the formula (4.61), which is expressed in a simpler mathematical form and which—as already mentioned—describes clothoid.

Table 4.3 Values of L [m] calculated on the basis of Eqs. (4.60) and (4.61)

Acceleration	Radius [m]	Speed					
		$v = 80$ km/h			$v = 120$ km/h		
According to Eq. (4.60)							
		$\sin\beta = 0.04$	$\sin\beta = 0$	$\sin\beta = -0.04$	$\sin\beta = 0.04$	$\sin\beta = 0$	$\sin\beta = -0.04$
1 m/s^2	4500	10.0	10.1	10.1			
	3000	15.2	15.3	15.4			
	2000	23.3	23.6	23.8			
	16000				9.4	9.4	9.4
	8000				18.9	19.0	19.1
	4500				34.3	34.5	34.7
−1 m/s^2	4500	9.4	9.5	9.5			
	3000	13.9	14.0	14.1			
	2000	20.4	20.6	20.8			
	16000				9.1	9.1	9.2
	8000				18.0	18.1	18.1
	4500				31.4	31.5	31.7
0 m/s^2	4500	9.7	9.8	9.8			
	3000	14.5	14.6	14.7			
	2000	21.8	22.0	22.1			
	16000				9.2	9.3	9.3
	8000				18.5	18.5	18.6
	4500				32.7	32.9	33.1
According to Eq. (4.60) (by $a = 1$ m/s^2 and without acceleration of gravity force)							
	4500	10.1					
	3000	15.3					
	2000	23.6					
	16000				9.4		
	8000				19.0		
	4500				34.5		
According to Eq. (4.61)							
	4500	9.8					
	3000	14.6					
	2000	21.9					
	16000				9.3		
	8000				18.5		
	4500				32.9		

4.3 General Transition Curves Described Using Curvature Function

Apart from the curves presented in Sects. 4.1 and 4.2, transition curves described using the curvature function include also so-called general transition curves (Grabowski 1984; Kobryń 2009). Their main characteristic is that a whole curvilinear transition between two straight lines is described using only one equation (Fig. 4.6). The general transition curve has only one maximum of the curvature, and values of the curvature at the start point P and end point K are equal to zero. First solutions of the general transition curves were presented by Grabowski (1984). A little later, considerations on this type transition curves were presented in papers (Baykal et al. 1998; Tari and Baykal 2005).

Different solutions of general transition curves are known. A basic equation, which is used for modeling the curvature using the appropriate boundary conditions has a form:

$$k(l) = \sum_{n=0}^{n=m} a_n l^n \tag{4.62}$$

The general transition curve which fulfills the conditions given by Eqs. (3.1), (3.7), (3.8) and (3.9) is described by the following equation (Grabowski 1984):

$$k(l) = \frac{1}{R_M} \left[A\frac{l}{L} + B\left(\frac{l}{L}\right)^2 + C\left(\frac{l}{L}\right)^3 \right] \tag{4.63}$$

where

$$A = \frac{2q - 3q^2}{q^2(1-q)^2}$$

$$B = \frac{-1 + 3q^2}{q^2(1-q)^2}$$

$$C = \frac{1 - 2q}{q^2(1-q)^2}$$

whereby:

$$q = l_M/L \left(\frac{1}{3} \leq q \leq \frac{2}{3}\right),$$

L total length of the curve,
l_M distance of the point M (in which it occurs a maximum curvature) from the initial point of the curve,
R_M radius of curvature at point M

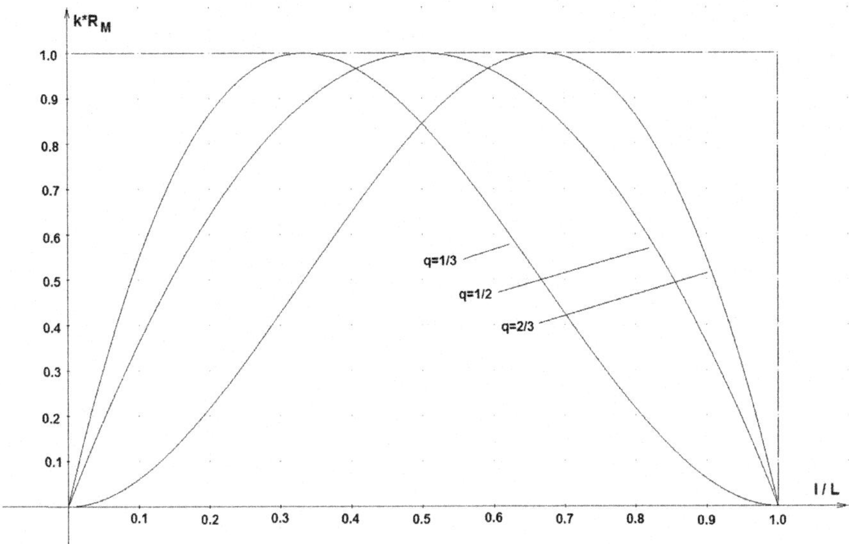

Fig. 4.6 Graphs of curvature for curves (4.63) by selected values q

According to (Kobryń 1993), the deflection angle at the point K is:

$$\hat{u} = \frac{L}{12R_M}\left(6A + 4B + 3C\right) \tag{4.64}$$

The general transition curve, which in addition to the conditions (3.1), (3.7), (3.8) and (3.9) also fulfills the conditions (3.3) and (3.4), has a form (Grabowski 1984):

$$k(l) = \frac{1}{R}\left[A\left(\frac{l}{L}\right)^2 + B\left(\frac{l}{L}\right)^3 + C\left(\frac{l}{L}\right)^4 + D\left(\frac{l}{L}\right)^5\right] \tag{4.65}$$

where:

$$A = \frac{3q - 8q^2 + 5q^3}{q^3(1-q)^4}$$

$$B = \frac{-2 + 12q^2 - 10q^3}{q^3(1-q)^4}$$

$$C = \frac{4 - 9q + 5q^3}{q^3(1-q)^4}$$

$$D = \frac{-2 + 6q - 4q^2}{q^3(1-q)^4}$$

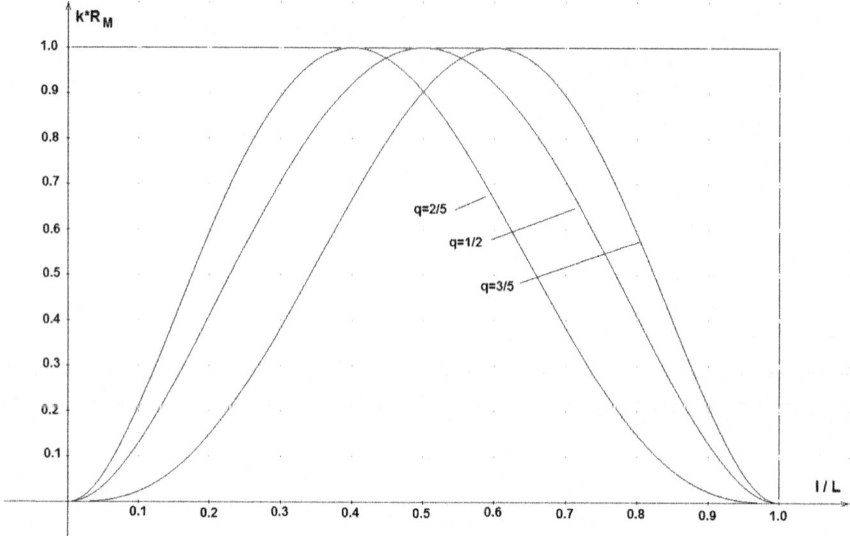

Fig. 4.7 Graphs of curvature for curves (4.65) by selected values q

whereby: $q = l_M/L$ and $\frac{4}{10} \leq q \leq \frac{6}{10}$.

Graphs of the curvature for the curves (4.62) and (4.64) for different values of the parameter q are shown in Figs. 4.6 and 4.7. They provide an illustration of high possibilities to shape the curvature of the curves (4.63) and (4.65). The result is great flexibility of adjustment the geometry of curvilinear transition to terrain limitations.

In paper (Kobryń 1993) two versions of formulas which describe the Cartesian coordinates of the curves (4.63) were presented. The first version takes the form:

$$x = l\left[1 - \left(\frac{l}{R_M}\right)^2 S_2 + \left(\frac{l}{R_M}\right)^4 S_4 - \left(\frac{l}{R_M}\right)^6 S_6 + \left(\frac{l}{R_M}\right)^8 S_8 - \cdots\right] \tag{4.66}$$

$$y = l\left[\frac{l}{R_M} S_1 - \left(\frac{l}{R_M}\right)^3 S_3 + \left(\frac{l}{R_M}\right)^5 S_5 - \left(\frac{l}{R_M}\right)^7 S_7 + \left(\frac{l}{R_M}\right)^9 S_9 - \cdots\right] \tag{4.67}$$

The second version of these formulas is as follows:

$$x = L\left[t - t^3 s^2 S_2 + t^5 s^4 S_4 - t^7 s^6 S_6 + t^9 s^8 S_8 - \cdots\right] \tag{4.68}$$

$$y = L\left[t^2 s S_1 - t^4 s^3 S_3 + t^6 s^5 S_5 - t^8 s^7 S_7 + t^{10} s^9 S_9 - \cdots\right] \tag{4.69}$$

wherein: $s = L/R_M$. Particular members S_i $(i = 1, 2, 3, \ldots, 10)$ occurring in formulas (4.66)–(4.69) have a following form:

$$S_1 = \frac{A}{6}t + \frac{B}{12}t^2 + \frac{C}{20}t^3$$

$$S_2 = \frac{A^2}{40}t^2 + \frac{AB}{36}t^3 + \frac{4B^2 + 9AC}{504}t^4 + \frac{BC}{96}t^5 + \frac{C^2}{288}t^6$$

$$S_3 = \frac{A^3}{336}t^3 + \frac{A^2B}{192}t^4 + \frac{8AB^2 + 9A^2C}{2592}t^5 + \frac{4B^3 + 27ABC}{6480}t^6$$
$$+ \frac{9AC^2 + 9B^2C}{6336}t^7 + \frac{BC^2}{1152}t^8 + \frac{C^3}{4992}t^9$$

$$S_4 = \frac{A^4}{3456}t^4 + \frac{A^3B}{1440}t^5 + \frac{3A^3C + 4A^2B^2}{6336}t^6 + \frac{8AB^3 + 27A^2BC}{31104}t^7$$
$$+ \frac{32B^4 + 243A^2C^2 + 432AB^2C}{808704}t^8 + \frac{8B^3C + 27ABC^2}{72576}t^9$$
$$+ \frac{3AC^3 + 4B^2C}{34560}t^{10} + \frac{BC^3}{18432}t^{11} + \frac{C^4}{104448}t^{12}$$

$$S_5 = \frac{A^5}{42240}t^5 + \frac{A^4B}{13824}t^6 + \frac{9A^4C + 16A^3B^2}{179712}t^7 + \frac{9A^3BC + 4A^2B^3}{72576}t^8$$
$$+ \frac{32AB^4 + 81A^3C^2 + 216A^2B^2C}{1866240}t^9 + \frac{32B^5 + 720AB^3C + 1215A^2BC^2}{14929920}t^{10}$$
$$+ \frac{32B^4C + 81A^2C^3 + 216AB^2C^2}{4230144}t^{11} + \frac{4B^3C^2 + 9ABC^2}{373248}t^{12}$$
$$+ \frac{9AC^4 + 16B^2C^3}{2101248}t^{13} + \frac{BC^4}{368640}t^{14} + \frac{C^5}{2580480}t^{15}$$

$$S_6 = \frac{A^6}{599040}t^6 + \frac{A^5B}{161280}t^7 + \frac{9A^5C + 20A^4B^2}{2073600}t^8 + \frac{27A^4BC + 16A^3B^3}{1990656}t^9$$
$$+ \frac{64A^2B^4 + 81A^4C^2 + 288A^3B^2C}{16920576}t^{10} + \frac{32AB^5 + 405A^3BC^2 + 360A^2B^3C}{33592320}t^{11}$$
$$+ \frac{128B^6 + 4320AB^4C + 3645A^3C^3 + 14580A^2B^2C^2}{19 \cdot 67184640}t^{12}$$
$$+ \frac{32B^5C + 360AB^3C^2 + 405A^2BC^3}{74649600}t^{13} + \frac{81A^2C^4 + 64B^4C^2 + 288AB^2C^3}{83607552}t^{14}$$
$$+ \frac{27ABC^4 + 16B^3C^3}{21897216}t^{15} + \frac{20B^2C^4 + 9AC^5}{50872320}t^{16} + \frac{BC^5}{8847360}t^{17} + \frac{C^6}{73728000}t^{18}$$

$$S_7 = \frac{A^7}{9676800}t^7 + \frac{A^6 B}{2211840}t^8 + \frac{3A^6 C + 8A^5 B^2}{9400320}t^9 + \frac{189A^5 BC + 140A^4 B^3}{2 \cdot 78382080}t^{10}$$

$$+ \frac{1701A^5 B^2 + 2240A^3 B^4 + 7560A^4 B^2 C}{57 \cdot 69672960}t^{11}$$

$$+ \frac{448A^2 B^5 + 2835A^4 BC^2 + 3360A^3 B^3 C}{25 \cdot 83607552}t^{12}$$

$$+ \frac{512B^7 + 3645A^4 C^3 + 8640A^2 B^4 C + 19440A^3 B^2 C^2}{120 \cdot 94058496}t^{13}$$

$$+ \frac{128B^7 + 6048AB^5 C + 25515A^3 BC^3 + 34020A^2 B^3 C^2}{440 \cdot 70543872}t^{14}$$

$$+ \frac{3645A^3 C^4 + 8640AB^4 C^2 + 19440A^2 B^2 C^3 + 512B^6 C}{288 \cdot 85847040}t^{15}$$

$$+ \frac{448B^5 C^2 + 2835A^2 BC^4 + 3360AB^3 C^3}{120 \cdot 83607552}t^{16}$$

$$+ \frac{243A^2 C^5 + 320B^4 C^3 + 1080AB^2 C^4}{125 \cdot 47775744}t^{17}$$

$$+ \frac{27ABC^5 + 20B^3 C^4}{13 \cdot 39813012}t^{18} + \frac{3AC^6 + 8B^2 C^5}{9 \cdot 53084160}t^{19} + \frac{BC^6}{4 \cdot 61931520}t^{20}$$

$$+ \frac{C^7}{29 \cdot 82575360}t^{21}$$

$$S_8 = \frac{A^8}{17 \cdot 10321920}t^8 + \frac{A^7 B}{34836480}t^9 + \frac{9A^7 C + 28A^6 B^2}{19 \cdot 23224320}t^{10}$$

$$+ \frac{567A^5 B^3 + 63A^6 BC}{10 \cdot 69672960}t^{11} + \frac{567A^6 C^2 + 1120A^4 B^4 + 3024A^5 B^2 C}{192 \cdot 91445760}t^{12}$$

$$+ \frac{448A^3 B^5 + 1701A^5 BC^2 + 2520A^4 B^3 C}{264 \cdot 52254720}t^{13}$$

$$+ \frac{1792A^2 B^6 + 5103A^5 C^3 + 20160A^3 B^4 C + 34020A^4 B^2 C^2}{1863 \cdot 92897280}t^{14}$$

$$+ \frac{512AB^7 + 12096A^2 B^5 C + 25515A^4 BC^3 + 45360A^3 B^3 C^2}{2916 \cdot 92897280}t^{15}$$

$$+ \frac{2048B^8 + 229635A^4 C^4 + 129024AB^6 C + 1088640A^2 B^4 C^2 + 1632960A^3 B^2 C^3}{145800 \cdot 92897280}t^{16}$$

$$+ \frac{512B^7 C + 12096AB^5 C^2 + 25515A^3 BC^4 + 45360A^2 B^3 C^3}{6318 \cdot 92897280}t^{17}$$

$$+ \frac{1792B^6 C^2 + 5103A^3 C^5 + 20160AB^4 C^3 + 34020A^2 B^2 C^4}{8748 \cdot 92897280}t^{18}$$

$$+ \frac{448B^5 C^3 + 1701A^2 BC^5 + 2520AB^3 C^4}{1512 \cdot 92897280}t^{19}$$

$$+ \frac{567A^2 C^6 + 1120B^4 C^4 + 3024AB^2 C^5}{4176 \cdot 92897280}t^{20}$$

$$+ \frac{8B^3 C^5 + 9ABC^6}{80 \cdot 59719680}t^{21} + \frac{9AC^7 + 28B^2 C^6}{496 \cdot 92897280}t^{22} + \frac{BC^7}{96 \cdot 82575360}t^{23}$$

$$+ \frac{C^8}{1056 \cdot 82575360}t^{24}$$

$$S_9 = \frac{A^9}{38 \cdot 92897280}t^9 + \frac{A^8 B}{10 \cdot 61931520}t^{10} + \frac{9A^8 C + 32A^7 B^2}{84 \cdot 92897280}t^{11}$$

$$+ \frac{28A^6 B^3 + 27A^7 BC}{48 \cdot 95800320}t^{12}$$

$$+ \frac{81A^7 C^2 + 224A^5 B^4 + 504A^6 B^2 C}{414 \cdot 92897280}t^{13} + \frac{224A^4 B^5 + 5675A^6 BC^2 + 1008A^5 B^3 C}{648 \cdot 92897280}t^{14}$$

$$+ \frac{3584A^3 B^6 + 5103A^6 C^3 + 4082A^5 B^2 C^2 + 30240A^4 B^4 C}{24300 \cdot 92897280}t^{15}$$

$$+ \frac{256A^2 B^7 + 5103A^5 BC^2 + 4032A^3 B^5 C + 11340A^4 B^3 C^2}{6318 \cdot 92897280}t^{16}$$

$$+ \frac{2048AB^8 + 45927A^5 C^4 + 64512A^2 B^6 C + 408240A^4 B^2 C^3 + 362880A^3 B^4 C^2}{314928 \cdot 92897280}t^{17}$$

$$+ \frac{2048B^9 + 165888AB^7 C + 1959552A^2 B^5 C^2 + 2066715A^4 BC^4 + 4898880A^3 B^3 C^3}{4408992 \cdot 92897280}t^{18}$$

$$+ \frac{2048B^8 C + 45927A^4 C^5 + 64512AB^6 C^2 + 362880A^2 B^4 C^3 + 408240A^3 B^2 C^4}{676512 \cdot 92897280}t^{19}$$

$$+ \frac{256B^7 C^2 + 4032AB^5 C^3 + 5103A^3 BC^5 + 11340A^2 B^3 C^4}{29160 \cdot 92897280}t^{20}$$

$$+ \frac{5103A^3 C^6 + 3584B^6 C^3 + 40824A^2 B^2 C^5 + 30240AB^4 C^4}{241056 \cdot 92897280}t^{21}$$

$$+ \frac{224B^5 C^4 + 567A^2 BC^6 + 1008AB^3 C^5}{13824 \cdot 92897280}t^{22} + \frac{81A^2 C^7 + 224B^4 C^5 + 504AB^2 C^6}{19008 \cdot 92897280}t^{23}$$

$$+ \frac{28B^3 C^6 + 27ABC^7}{4896 \cdot 92897280}t^{24} + \frac{9AC^8 + 32B^2 C^7}{17920 \cdot 92897280}t^{25} + \frac{BC^8}{3072 \cdot 92897280}t^{26}$$

$$+ \frac{C^9}{37888 \cdot 92897280}t^{27}$$

As can be seen, formulas (4.66)–(4.69) are very complex, which should be considered as a significant drawback of curves described using the curvature function. Due to the higher degree of the curve equation, formulas describing the coordinates for the curve (4.65) would be even more complex. Alternatively, the coordinates can be calculated using the integrals (3.10) and (3.11) with different numerical integration methods.

References

Auberlen R (1956) Vom Schwung der Fahrt zur Form der Strasse. Forschungsarbeiten aus dem Straßenwesen, Heft 25, Bielefeld (in German)

Baykal O, Tari E, Coskun Z, Sahin M (1998) New transition curve joining two straight lines. J Transp Eng 124(5):337–345

Durth W (1974) Ein Beitrag zur Erweiterung des Modells für Fahrer, Fahrzeug und Straße in der Straßenplanung. Schriftenreihe Straßenbau und Straßenverkehrstechnik, Heft 163. Hrsg. vom Bundesminister für Verkehr, Bonn (in German)

Grabowski RJ (1973) Optymalizacja krzywych przejściowych przystosowanych do dużych prędkości. Drogownictwo 28(3):70–76 (in Polish)

Grabowski RJ (1984) Gładkie przejścia krzywoliniowe w drogach kołowych i kolejowych. Zeszyty Naukowe AGH, Geodezja nr 82, Kraków (in Polish)

Grabowski RJ, Kobryń A (1997) Krzywe przejściowe z gładkim wykresem krzywizny jako element geometrii trasy drogowej. Drogownictwo 52(7):212–214 (in Polish)

Klein R (1937) Beitrag zur Gestaltung der Übergangsbögen. Gleistechnik und Fahrbahnbau, Heft 13 (in German)

Kobryń A (1991a) Zur Berechnung von Absteckdaten bei der Anwendung der Bloßkurve. Zeitschrift für Vermessungswesen 116(7):284–288 (in German)

Kobryń A (1991b) Zur Kurvenüberleitung bei der Ausfahrt von Autobahnen. Vermessungs-wesen und Raumordnung 53(8):385–392 (in German)

Kobryń A (1993) Allgemeine mathematische Übergangskurven als Trassierungselement. Zeitschrift für Vermessungswesen 118(5):227–242 (in German)

Kobryń A (1995) Krzywe hamowania w trasowaniu węzłów drogowych. Drogownictwo 50 (1).12–14 (in Polish)

Kobryń A (1999) Geometryczne kształtowanie krzywoliniowych odcinków niwelety tras drogowych. Wydawnictwa Politechniki Białostockiej, Rozprawy Naukowe nr 60, Białystok (in Polish)

Kobryń A (2007) Krzywe przejściowe na pionowych łukach tras drogowych w świetle warunków dynamiki ruchu. Drogownictwo 62(9):291–295 (in Polish)

Kobryń A (2008) Klotoida i inne krzywe przejściowe. Drogownictwo 63(6–7):189–195 (in Polish)

Kobryń A (2009) Wielomianowe kształtowanie krzywych przejściowych. Wydawnictwa Politechniki Białostockiej, Rozprawy Naukowe nr 167, Białystok (in Polish)

Lamm R, Psarianos B, Mailänder T (1999) Highway design and traffic safety engineering handbook. McGraw-Hill, Professional Book Group, New York

Leyko J (1996) Mechanika ogólna. Tom 1, 2. Wydawnictwo Naukowe PWN, Warszawa (in Polish)

Lipiński M (1993) Geometria i tyczenie tras drogowych. In: Geodezja inżynieryjna, tom 3. Polskie Przedsiębiorstwo Wydawnictw Kartograficznych, Warszawa – Wrocław (in Polish)

Lorenz H (1971) Trassierung und Gestaltung vion Strassen und Autobahnen. Wiesbaden – Berlin (in German)

McConell WA (1957) Human sensitivity to motion as a design criterion for highway curves. Bulletin No. 149, Highway Research Board, Washington

Melchior P (1928) Der Ruck. VDI - Zeitschrift 72, Heft 50 (in German)

Meyer CF, Gibson DW (1980) Route surveying and design. Harper & Row, New York

Tari E, Baykal O (2005) A new transition curie with enhanced properties. Can J Civ Eng 32 (5):913–923

Chapter 5
Transition Curves Described Using Explicit Function

Among different transition curves that are described using the curvature function can be mentioned different solutions that may be useful in various applications related to the geometrical shaping of highways. So-called polynomial transition curves, which will be presented in Chap. 7, occupy a special place here. Whereas, in this section two solutions: parabolic transition curves and sinusoidal transition curves are described.

5.1 Parabolic Transition Curves

Parabolic transition curves were presented in the work (Kobryń 2009). The equation of these curves was defined based on the function in the form:

$$y = ax^n. \tag{5.1}$$

Within the graph of the function (5.1) a fragment, which corresponds to the classical definition of the transition curves (bold part of the graph on Fig. 5.1), can be identified.

To identify this fragment, it was assumed that a tangent inclination at the end point K with abscissa $x = x_K$ should be equal to $\tan u_K$. Since $y' = \tan u_K$, from (5.1) it follows:

$$\tan u_K = nax_K^{n-1} \tag{5.2}$$

It follows that:

$$a = \frac{\tan u_K}{nx_K^{n-1}}. \tag{5.3}$$

© Springer International Publishing AG 2017

A. Kobryń, *Transition Curves for Highway Geometric Design*, Springer Tracts on Transportation and Traffic 14, DOI 10.1007/978-3-319-53727-6_5

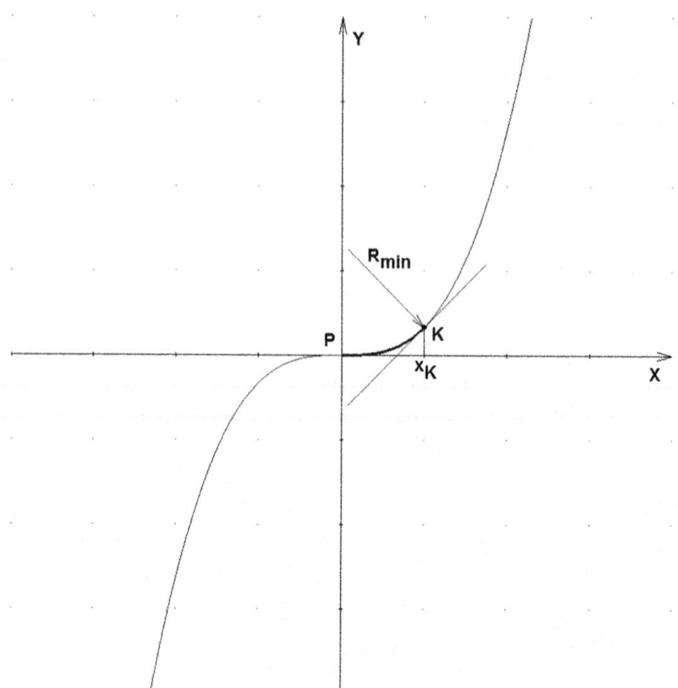

Fig. 5.1 Graph of the parabola n-th degree and her section as a transition curve

Therefore, can be written Eq. (5.1) as:

$$y = \frac{x_K \tan u_K}{n} t^n \tag{5.4}$$

where: $t = x/x_K$, $x \in \langle 0; x_K \rangle$, $n \in N$.

Classically understood transition curves should provide a gradual increase in the curvature from a minimum value to a specified maximum value $1/R_{min}$, where R_{min} is the minimum radius of curvature. Derivatives y' and y'' of the function (5.4) are:

$$y' = \tan u_K \frac{x^{n-1}}{x_K^{n-1}}, \tag{5.5}$$

$$y'' = \frac{(n-1)\tan u_K}{x_K} \frac{x^{n-2}}{x_K^{n-2}} \tag{5.6}$$

Therefore, from Eq. (3.13) it follows, that the curvature of the curve (5.4) in the range of abscissa $x \in \langle 0; x_K \rangle$ can be expressed as:

$$k = \frac{(n-1)\cdot \tan u_K \cdot x_K^{2n-2} x^{n-2}}{\left(x_K^{2n-2} + \tan^2 u_K \cdot x^{2n-2}\right)^{3/2}} \tag{5.7}$$

From (3.7) follows that for $n = 1$ and $n > 2$ the curvature at point $P(x = 0)$ is zero. Nonzero curvature at this point occurs if $n = 2$, whereby from Eq. (5.7) results that the curvature is equal to $k = \frac{\tan u_K}{x_K}$. It follows that the curves (5.4) have a curvature graph consistent with a definition of the classically understood transition curves when $n > 2$. The curvature has a maximum at a point, which abscissa depends on the abscissa x and the value of the parameter a is specified by the Eq. (5.3). Based on the necessary condition (3.21) for the curves (5.4) it follows

$$\frac{dk}{dx} = \frac{(n-1)x_K^{2n-2}\cdot \tan u_K \left[x_K^{2n-2}x^{n-3} - (2n-1)\tan u_K \cdot x^{3n-5}\right]}{\left(x_K^{2n-2} + \tan^2 u_K \cdot x^{2n-2}\right)^{5/2}} \tag{5.8}$$

Thus, the necessary condition of existence an extremum requires that:

$$x_K^{2n-2}x^{n-3} - (2n-1)\tan^2 u_K \cdot x^{3n-5} = 0 \tag{5.9}$$

Assuming that the curvature should reach an extremum for $x = x_K$, from Eq. (5.9) follows:

$$x_K^{3n-5} = (2n-1)\tan^2 u_K \cdot x_K^{3n-5} \tag{5.10}$$

Finally, it is obtained:

$$\tan u_K = \sqrt{\frac{1}{2n-1}}. \tag{5.11}$$

Eq. (5.11) determines the maximal values of $\tan u_K$ for the parabola of n-th degree (by $n > 2$), for which a beginning part in the range $x \in \langle 0; x_K \rangle$ can be used as the transition curve. Function (5.7), which describes the curvature, will be growing in the entire range $x \in \langle 0; x_K \rangle$, if the condition $\frac{dk}{dx} > 0$ will be fulfilled. Because $\tan u_K > 0$, therefore it should be:

$$x^{n-3}\left[x_K^{2n-2} - (2n-1)\tan^2 u_K \cdot x^{2n-2}\right] > 0. \tag{5.12}$$

Because $x \leq x_K$ and $2n - 1 > 0$, $\tan u_K > 0$, inequality (5.12) will be fulfilled if

$$x_K^{2n-2} - (2n-1)\tan^2 u_K \cdot x_K^{2n-2} > 0 \tag{5.13}$$

whence follows

$$1 - (2n - 1)\tan^2 u_K > 0 \tag{5.14}$$

It follows:

$$\tan u_K < \sqrt{\frac{1}{2n - 1}}. \tag{5.15}$$

When $n > 2$, conditions (5.11) and (5.15) can be written as

$$\tan u_K \leq \sqrt{\frac{1}{2n - 1}}. \tag{5.16}$$

The curvature described by Eq. (5.7) should not exceed the designed maximum value. To do this, it should be determined a minimal length of the abscissa x_K, which corresponds to assumed values R_K and $\tan u_K$. Assuming that for $x = x_K$ the curvature should be $k = 1/R_K$, from Eq. (5.7) implies a following equation:

$$\frac{x_K}{R_K} = \frac{(n - 1)\tan u_K}{(1 + \tan^2 u_K)^{3/2}} \tag{5.17}$$

Equation (5.17) can also be written as:

$$x_K = R_K \frac{(n - 1)\tan u_K}{(1 + \tan^2 u_K)^{3/2}} \tag{5.18}$$

From Eqs. (5.4) and (5.18) it follows that an ordinate of the point K is:

$$y_K = R_K \frac{n - 1}{n} \frac{\tan^2 u_K}{(1 + \tan^2 u_K)^{3/2}} \tag{5.19}$$

5.2   Sinusoid

Sinusoid has been described in the literature as a transition curve primarily for applications in railway (e.g. Bałuch 1983). The equations allowing convenient use of sinusoid also in road applications are given below. A generalized equation of the sinusoid has a form:

$$y = a \sin \omega \tag{5.20}$$

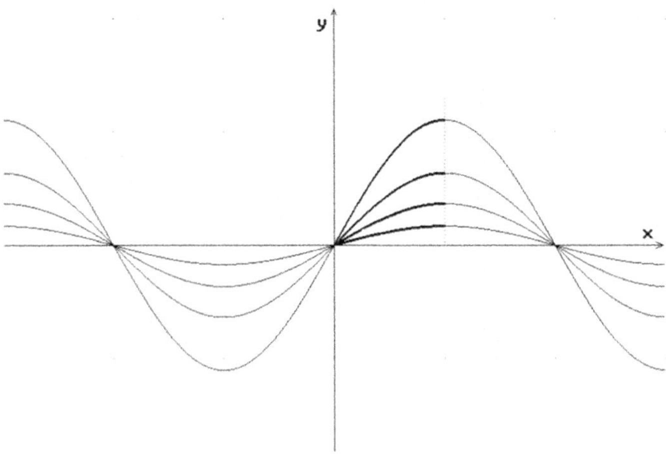

Fig. 5.2 Sinusoid as transition curve

The paper (Kobryń 2006) examines a part of the sinusoid for the angle $\omega \in \langle 0; \Pi/2 \rangle$ (Fig. 5.2). It was assumed that the end point K of the transition curve, which is formed by an arc of the sinusoid in the given range of angles ω, corresponds to the abscissa x_K. The point K with the abscissa x_K corresponds to the angle equal to $\omega = \Pi/2$. This is equivalent to the condition:

$$\frac{x}{x_K} = \frac{\omega}{\Pi/2} \tag{5.21}$$

where x is an abscissa of any point of the sinusoid located between the start point P and end point K. The value x corresponds to the angle $\omega \in \langle 0; \Pi/2 \rangle$.

From Eq. (5.21) follows:

$$\omega = \frac{\Pi}{2} \frac{x}{x_K} \tag{5.22}$$

Substituting it into Eq. (5.20) is obtained:

$$y = a \, \sin\left(\frac{\Pi}{2} \frac{x}{x_K}\right) \tag{5.23}$$

Derivative y' of function (5.23) has a following form:

$$y' = a \frac{\Pi}{2x_K} \cos\left(\frac{\Pi}{2} \frac{x}{x_K}\right) \tag{5.24}$$

At the initial point $P(x = 0)$ should be $y' = \tan u_P$, whereby $\tan u_P > 0$. It follows:

$$\tan u_P = a\frac{\Pi}{2x_K} \tag{5.25}$$

and:

$$a = \frac{2x_K \tan u_P}{\Pi} \tag{5.26}$$

Taking into account Eq. (5.26), Eq. (5.21) can be finally written as:

$$y = \frac{2x_K \tan u_P}{\Pi}\sin\left(\frac{\Pi}{2}\frac{x}{x_K}\right) \tag{5.27}$$

From Eq. (5.27) for $x = x_K$ it follows that for the point K is:

$$y_K = \frac{2x_K \tan u_P}{\Pi} \tag{5.28}$$

Derivative y' of function (5.27) can be written as:

$$y' = \tan u_P \cos\left(\frac{\Pi}{2}\frac{x}{x_K}\right) \tag{5.29}$$

From Eq. (5.29) it follows that for $x = 0$ is $y' = \tan u_P$, whereas for $x = x_K$ is $y' = 0$.

At the point K with abscissa x_K (corresponding to the angle $\omega = \Pi/2$) the curvature should reach an extreme value equal to $-1/R_K$. On the basis (3.21), the necessary condition for extremum of the curvature can be written as:

$$-\frac{\Pi}{4x_K^2}\cos\left(\frac{\Pi}{2}t\right)\left[\operatorname{tg} u_P\left(1+\operatorname{tg}^2 u_P \cos^2\left(\frac{\Pi}{2}t\right)\right)+3\operatorname{tg}^3 u_P \sin^2\left(\frac{\Pi}{2}t\right)\right] = 0 \tag{5.30}$$

whereby $t = x/x_K$, derivative y' is described by Eq. (5.29), and derivatives y'' and y''' are:

$$y'' = -\frac{\Pi^2}{2x_K}\operatorname{tg} u_P \sin\left(\frac{\Pi}{2}\frac{x}{x_K}\right) \tag{5.31}$$

and

$$y''' = -\frac{\Pi^2}{4x_K^2}\,\mathrm{tg}\,u_P\cos\left(\frac{\Pi}{2}\frac{x}{x_K}\right) \tag{5.32}$$

According to analyzes described in article (Kobryń 2006), the condition (5.30) is fulfilled when:

$$-\frac{1}{R_K} = -\frac{\Pi}{2x_K}\tan u_P\sin\left(\frac{\Pi}{2}\frac{x}{x_K}\right) \tag{5.33}$$

From (5.33) it follows

$$\frac{R_K\tan u_P}{x_K} = \frac{2}{\Pi} \tag{5.34}$$

The formula (5.34) is a main design condition of the sinusoid as the transition curve and describes a relationships between R_K, x_K and $\tan u_P$.

The paper (Kobryń 2006) also contains other relevant formulas which relate to the sinusoid defined by Eq. (5.25). On the basis of Eq. (5.34) the abscissa x_K can be expressed as:

$$x_K = R_K\frac{\Pi}{2}\tan u_P \tag{5.35}$$

Substituting (5.35) to (5.28) an ordinate is obtained:

$$y_K = R_K\tan^2 u_P \tag{5.36}$$

Since $\tan\alpha = y_K/x_K$, so on the basis (5.35) and (5.36) it follows:

$$\tan u_P = \frac{\Pi}{2}\tan\alpha \tag{5.37}$$

On the basis of Eqs. (5.35) and (5.36) can be also specified a formula, which determines the length of the chord connecting the points P and K:

$$s = \frac{R_K\tan u_P}{2}\sqrt{\Pi^2 + 4\tan^2 u_P} \tag{5.38}$$

In addition to the general characteristics of a sinusoid as the transition curie, graphs of the curvature were presented on Fig. 5.3 (for angles u_P equal to $15°$, $30°$, $45°$ and $60°$).

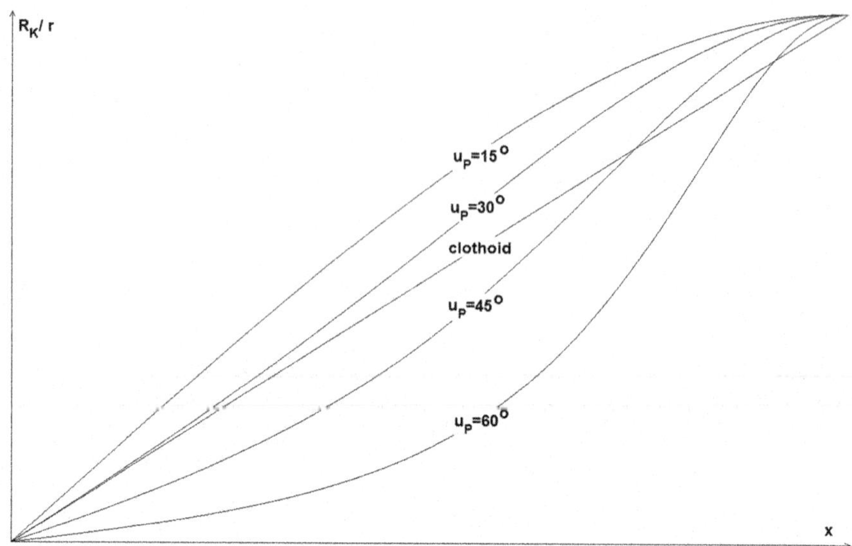

Fig. 5.3 Example graphs of curvature for sinusoid

Sections of the sinusoid graph in the range of angles $\omega \in \langle 0; \Pi \rangle$ can be used as general transitions curves (according to the definition of these curves given in the Chap. 3). The equation of this curve can be obtained on the basis of (5.20) for angles $\omega \in \langle 0; \Pi \rangle$.

Similarly, as in previous studies in this section, the initial equation of the sinusoid as the general transition curve can be written as:

$$y = a \, \sin\left(\Pi \frac{x}{x_K} \right) \qquad (5.39)$$

Derivative y' of function (5.39) is:

$$y' = a \frac{\Pi}{x_K} \cos\left(\Pi \frac{x}{x_K} \right) \qquad (5.40)$$

Tangent inclinations at the start point P and end point K are denoted respectively as $\tan u_P$ and $\tan u_K$. As a result, on the basis of Eq. (5.40) follows:

- for $x = 0$

$$a = \frac{x_K \tan u_P}{\Pi} \qquad (5.41)$$

- for $x = x_K$

$$a = -\frac{x_K \tan u_P}{\Pi} \qquad (5.42)$$

From comparison the above equations it follows $\tan u_P = \tan u_K$. If we denote $\tan u_P = -\tan u_K = \tan \hat{u}$, than Eq. (5.39) takes a form:

$$y = \frac{x_K \tan \hat{u}}{\Pi} \sin\left(\Pi \frac{x}{x_K}\right) \qquad (5.43)$$

Eq. (5.43) describes the sinusoid as the general transition curve. On the basis of Eq. (3.21) it can be stated that the maximal curvature exists in the case of the function (5.43) for $t = 1/2$, i.e. for $x = 0.5 \cdot x_K$. Denoting the radius of curvature at this point as R_S, on the basis of Eq. (3.13) can be stated that the following relation is fulfilled in the case of the sinusoid as the general transition curve (with a symmetrical distribution of curvature):

$$\frac{R_S \tan \hat{u}}{x_K} = \frac{1}{\Pi} \qquad (5.44)$$

References

Bałuch H (1983) Optymalizacja układów geometrycznych toru. Wydawnictwa Komunikacji i Łączności, Warszawa (in Polish)
Kobryń A (2006) Sinusoida jako krzywa przejściowa. Drogownictwo 61(7):242–245 (in Polish)
Kobryń A (2009) Wielomianowe kształtowanie krzywych przejściowych. Wydawnictwa Politechniki Białostockiej, Rozprawy Naukowe nr 167, Białystok (in Polish)

Chapter 6
Transition Curves Defined in the Polar Coordinate System

The lemniscate of Bernoulli may be mentioned especially in this group of transition curves. This curve is ideal for routing of ramps within the road interchanges. Therefore, it is advisable to locate the lemniscate in an orthogonal coordinate system as shown in Fig. 6.1 (Grabowski 1996).

Regardless of the coordinate system, so-called natural equation of the lemniscate has a following form:

$$r \cdot \rho = \frac{a^2}{3} = const.$$
(6.1)

wherein:

r the radius of curvature at any point i,
ρ the chord, which is the radius-vector O-i,
a parameter equal to the longest chord OM, that is $a = \rho_{max}$.

From Eq. (6.1) it follows that the lemniscate curvature is proportional to the radius-vector (i.e. to the chord).

An equation of the lemniscate on the polar form is as follows (Grabowski 1996):

$$\rho^2 = a^2 \sin 2\omega$$
(6.2)

The radius of curvature at any point is:

$$r = \frac{a}{3\sqrt{\sin 2\omega}} = \frac{a^2}{3\rho}$$
(6.3)

© Springer International Publishing AG 2017
A. Kobryń, *Transition Curves for Highway Geometric Design*, Springer Tracts
on Transportation and Traffic 14, DOI 10.1007/978-3-319-53727-6_6

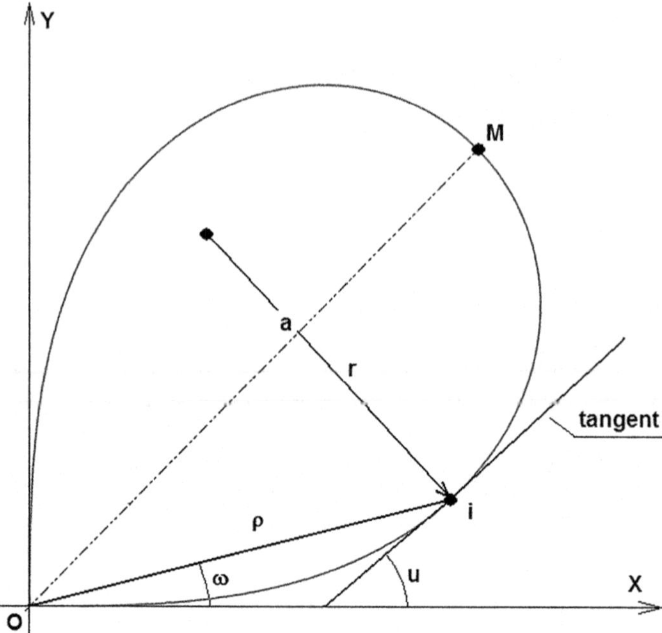

Fig. 6.1 Lemniscate in the Cartesian coordinate system

From Eq. (6.3) it follows that the radius at the point O (for $\omega = 0$) is $r = \infty$, and then decreases with increasing of the angle ω. The radius reaches a minimum value at the point M, that is, for $\omega = 45°$. It is equals:

$$r_{\min} = \frac{a}{3} \tag{6.4}$$

From Fig. 6.1 it follows that rectangular coordinates (related to tangent) can be written as:

$$x = \rho \cos \omega = a \cos \omega \sqrt{\sin 2\omega} \tag{6.5}$$

$$x = \rho \sin \omega = a \sin \omega \sqrt{\sin 2\omega} \tag{6.6}$$

The deflection angle u is always three times greater than the polar angle ω for a given point, i.e.:

$$u = 3\omega \tag{6.7}$$

It follows that the angle between the tangent and chord at any point is always 2ω. This is an important rule that can be used in a setting out a curve. It should be added that the condition (6.7) is approximately fulfilled also along some stretch of the clothoid (Lipiński 1993).

As seen, all formulas describing lemniscate are very simple. Only the calculation of the arc length is more complicated because it requires the use of elliptic integrals in the form:

$$L = a \int_0^\omega \frac{d\omega}{\sqrt{\sin 2\omega}} \tag{6.8}$$

References

Grabowski RJ (1996) Kształtowanie geometryczne krzywych przejściowych w drogach kołowych, kolejowych i trasach wodnych. Wydawnictwa Politechniki Białostockiej, Rozprawy Naukowe nr 38, Białystok (in Polish)

Lipiński M (1993) Geometria i tyczenie tras drogowych. In: Geodezja inżynieryjna, tom 3. Polskie Przedsiębiorstwo Wydawnictw Kartograficznych, Warszawa – Wrocław (in Polish)

Chapter 7
Polynomial Description of Transition Curves

Using an explicit function for description of transition curves is very comfortable in the light of laying out of these geometric elements. One of these forms of description of transition curves is to use polynomial functions. This chapter presents different solutions of polynomial transition curves.

7.1 Categories of Polynomial Transition Curves

Various solutions of transition curves, which were determined on the basis of polynomial functions $y = f(x)$ with appropriate design assumptions with respect to the tangent inclination and curvature, were presented in the work Kobryń (2009). The aim of those research was to develop a universal set of geometric tools that could be used for solving of various design problems in the setting out of horizontal and vertical alignment of roads and highways.

The proposed classification of transition curves includes following categories:

- transition curves in a classic sense
- quasi-transition curves
- general transition curves
- universal transition curves

Apart from the above categories, it can be identified two additional:

- S-shaped transition curves
- oval transition curves

According to the classical understanding, as the **transition curves** should be understood such curves which provide a gradual increase of curvature from zero at the start point to a certain maximum value at the end point. Depending on assumed location of the curves in the local coordinate system, it can be identified the curves

© Springer International Publishing AG 2017

A. Kobryń, *Transition Curves for Highway Geometric Design*, Springer Tracts on Transportation and Traffic 14, DOI 10.1007/978-3-319-53727-6_7

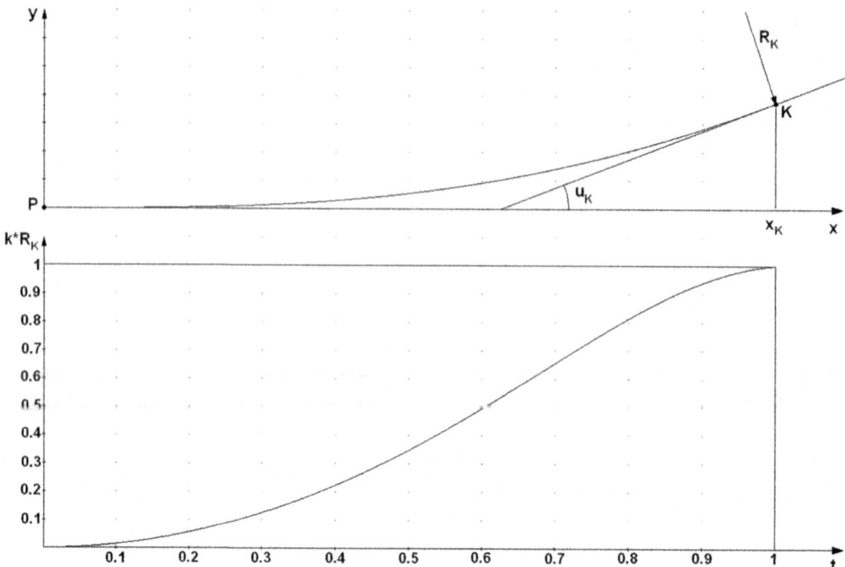

Fig. 7.1 Transition curve with a horizontal tangent at start point and graph of her curvature

with a horizontal tangent at the start point P (Fig. 7.1) and the curves with a horizontal tangent at the end point K (Fig. 7.2).

As quasi-transition curves are understood curves which have one maximum of curvature, but not at the end point, but at some point within the curve (Figs. 7.3 and 7.4). Such curves could be used in the design of curvilinear transition under various terrain restrictions, giving an opportunity to obtain an another location of route. Similarly to conventionally understood transition curves, for the quasi-transition curves can be assumed a distribution into two groups: the curves with a horizontal tangent at the starting point P (Fig. 7.3) and the curves with the horizontal tangent at the end point K (Fig. 7.4).

The notion of **general transition curves** were used in the work Grabowski (1984) in relation to the curves, within whose a curvature is increasing from zero at the starting point to a specified maximum value, then decreases to zero at the end point (Fig. 7.5). In the case of the general transition curves, the whole curvilinear transition between the two straight lines is described by only one equation.

As **S-shaped transition curves** should be understood the curves that the curvature at the start and end is zero, but in addition they have a inflection point (Fig. 7.6). The inflection point divides the curve into two parts, each of them has a curvature distribution appropriate for the general transition curves.

As **universal transition curves** should be understood the curves that provide a continuous change of curvature within a whole curvilinear transition between any two points, which are characterized by any values and signs of curvature and any deflection angles of tangents. Furthermore, such curves have one inflection point (Fig. 7.7).

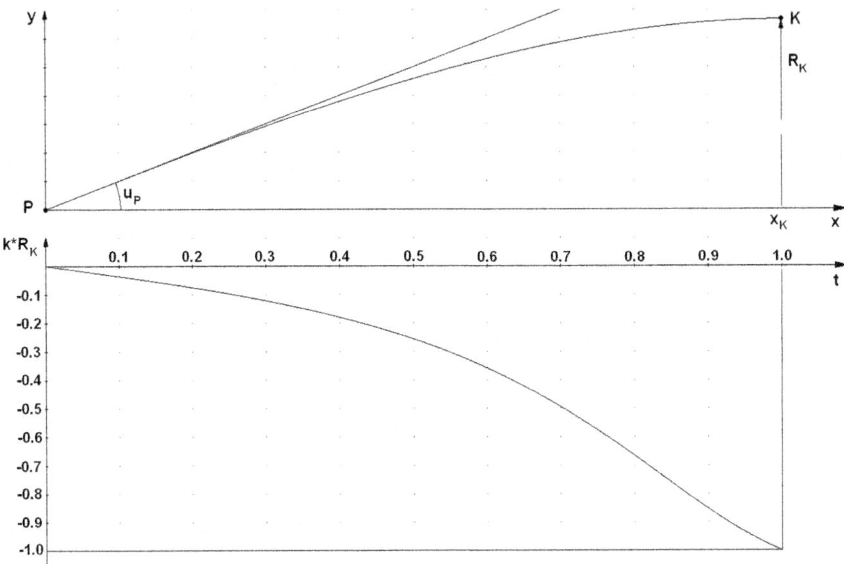

Fig. 7.2 Transition curve with a horizontal tangent at end point and graph of her curvature

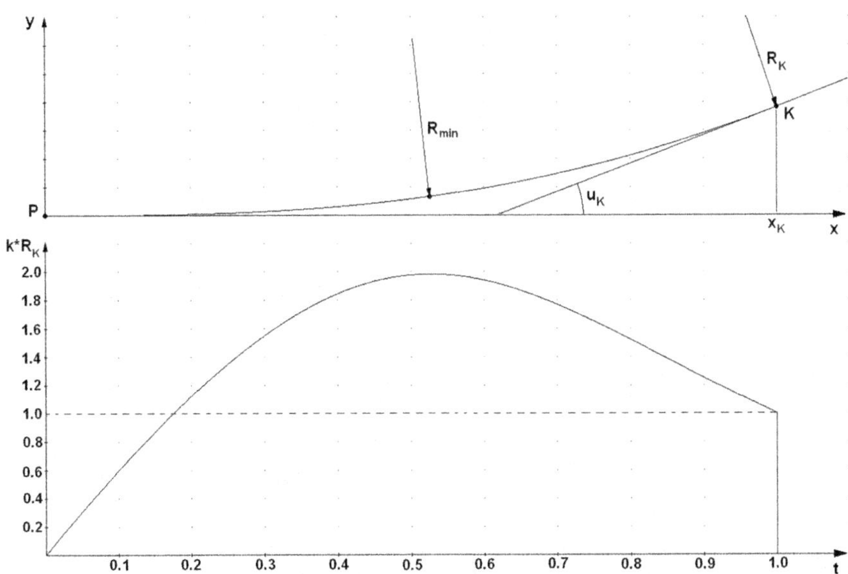

Fig. 7.3 Quasi-transition curve with a horizontal tangent at start point and graph of her curvature

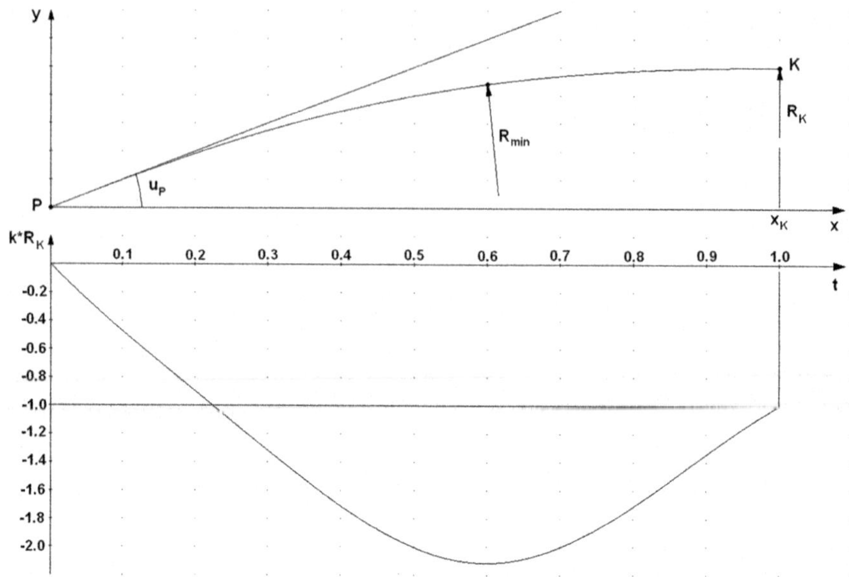

Fig. 7.4 Quasi-transition curve with a horizontal tangent at end point and graph of her curvature

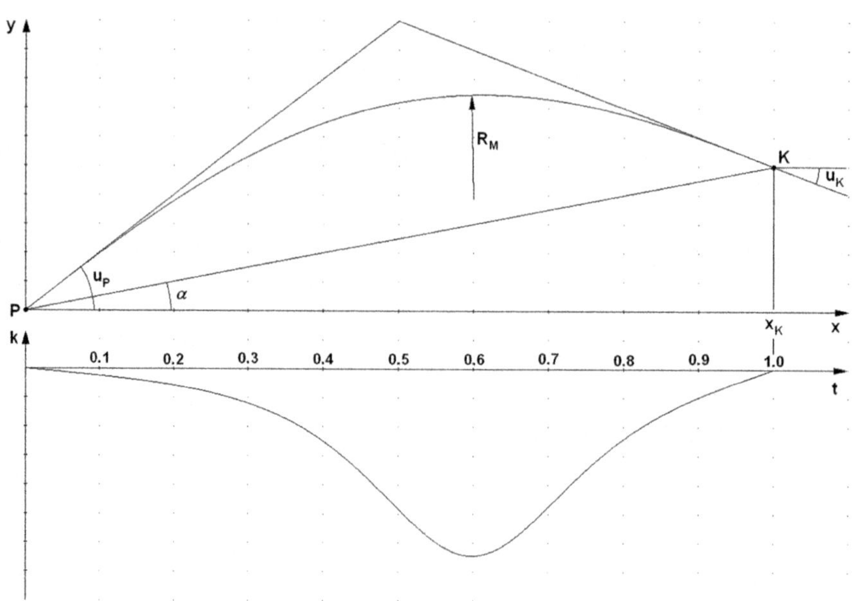

Fig. 7.5 General transition curve and graph of her curvature

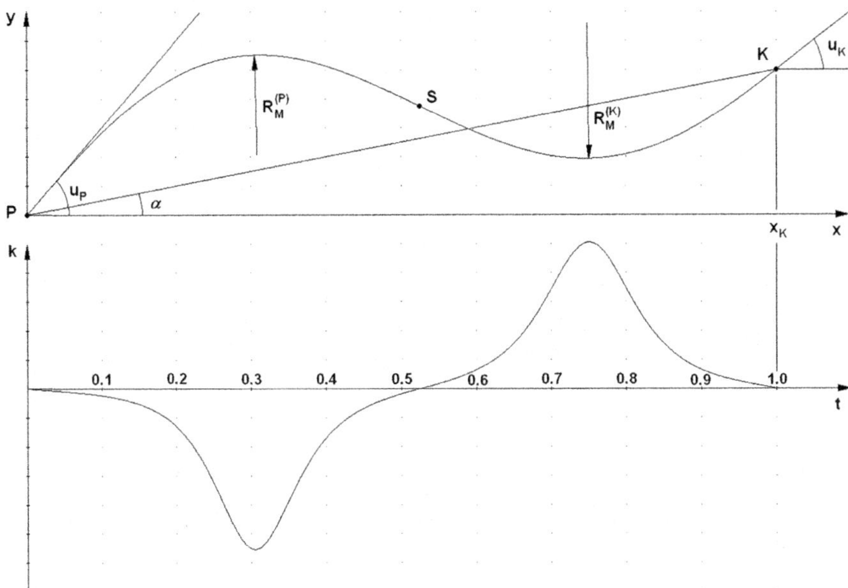

Fig. 7.6 S-shaped transition curve and graph of her curvature (*with permission from ASCE*)

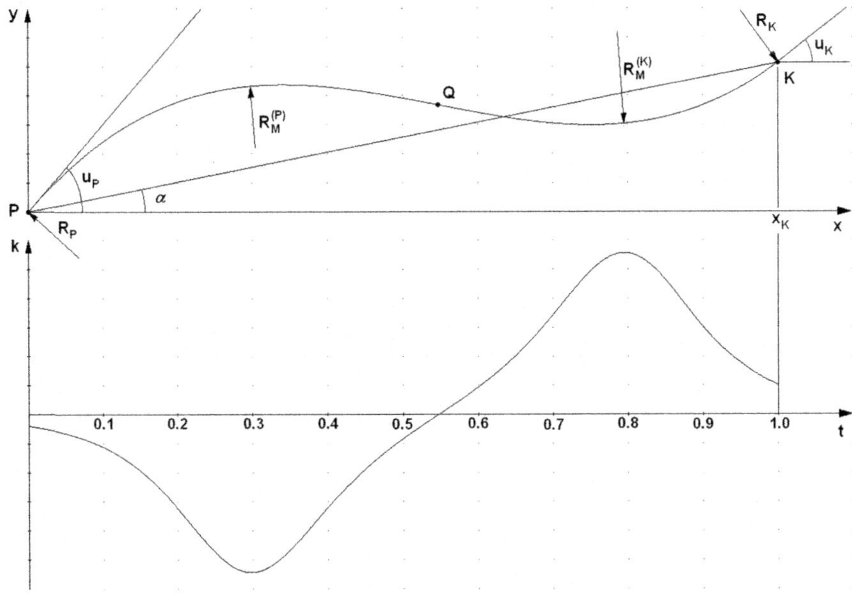

Fig. 7.7 Universal transition curve and graph of her curvature (*with permission from ASCE*)

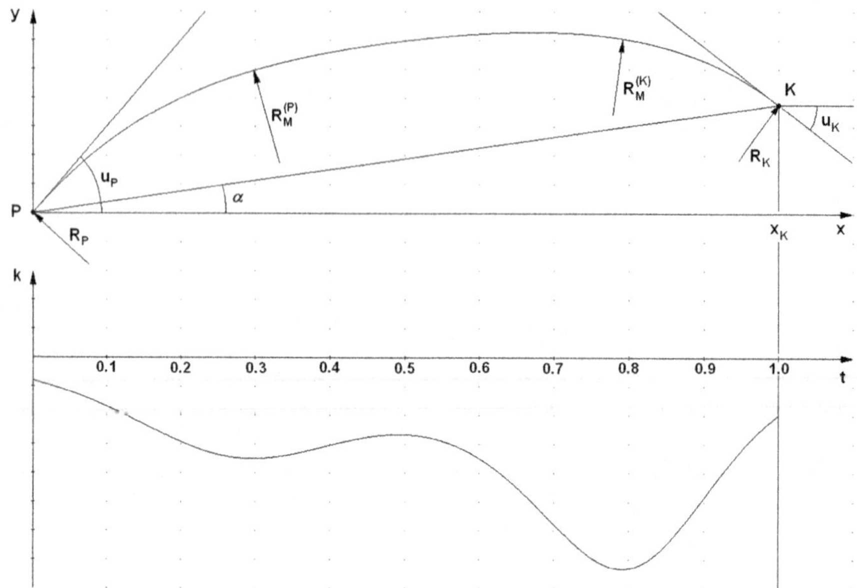

Fig. 7.8 Oval transition curve and graph of her curvature (*with permission from ASCE*)

Whereas as the **oval transition curves** should be understood a such curves that also provide a continuous change of curvature along the whole curvilinear transition between any two points, but do not have an inflection point (Fig. 7.8).

7.2 Boundary Conditions for Polynomial Transition Curves

The basis for the identification of the appropriate polynomial transition curves is a polynomial function in the form:

$$y = f(x) = \sum_{i=0}^{i=n} a_i x^i \tag{7.1}$$

Obtaining the correct distribution of curvature within the designed transition curve is mainly associated with the location of the point at which curvature is maximal. Additionally, the curve geometry is affected by assumed tangent inclinations to the curve graph. Other parameters can also play a supportive role in designing the geometry of the transition curve. In the case of curves that are defines using the function $y = f(x)$, their curvature is described by the equation:

$$k = \frac{y''}{\left(1 + y'^2\right)^{3/2}} \tag{7.2}$$

A location of the curvature extremum for the curve described by Eq. (7.2) corresponds to zero values of a derivative:

$$\frac{dk}{dx} = \frac{y'''\left(1 + y'^2\right) - 3y'y''^2}{\left(1 + y'^2\right)^{5/2}} \tag{7.3}$$

that is results from an equation:

$$y'''\left(1 + y'^2\right) - 3y'y''^2 = 0 \tag{7.4}$$

Assuming certain assumptions regarding the values of ordinates y and derivatives, y', y'' and y''', on the basis of the function (7.1) appropriate solutions of polynomial transition curves can be obtained. These assumptions are used as the boundary conditions to define appropriate solutions of polynomial transition curves. These conditions are primarily:

- values of the ordinates y_P and y_K,
- values of the tangent inclinations y'_P and y'_K,
- values of the second derivatives y''_P and y''_K,

According to Eq. (7.2), values of the derivatives y''_P and y''_K, including values of the derivatives y'_P and y'_K, determine the values of the curvature at points P and K.

In addition, for shaping the geometry of polynomial transition curves values of the derivatives y'''_P and y'''_K can be used. In certain circumstances they may define the smoothness of the curvature graph at the point P and K (depending on values of the derivatives y', y'' at these points). According to the Figs. 7.1, 7.2, 7.3, 7.4, 7.5, 7.6, 7.7 and 7.8, abscissa of the point P is $x = 0$. In addition, it should be assumed that the abscissa of the point K is $x = x_K$.

7.3   Generalized Solutions of Polynomial Transition Curves

In the work Kobryń (2009) were presented generalized equations of the polynomial transition curves, which were determined on the basis of the function (7.1). The generalized character of these solutions follows from possibility of adopting any detailed assumptions regarding the values y_P, y_K, y'_P, y'_K, y''_P, y''_K, y'''_P and y'''_K. Taking into account detailed assumptions in relation to these values and based on generalized solutions, appropriate equations of transition curves in the various categories can be determined, are defined in Sect. 7.1.

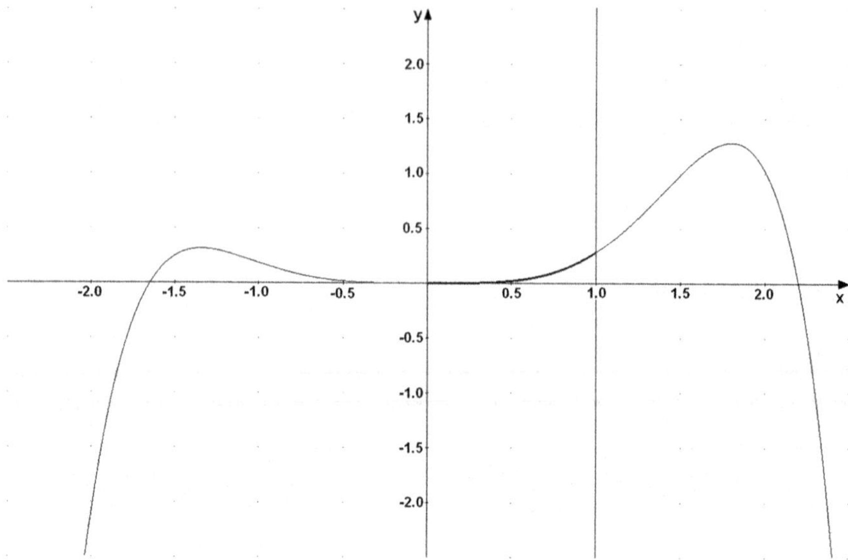

Fig. 7.9 Illustration a result of replacing the variable x by variable t

In any of the generalized solution, the variable x is replaced by the variable $t = x/x_K$, which may take the values $t \in \langle 0; 1 \rangle$. That assumption is convenient for practical reasons. It provides the standardization of appropriate solutions of the transition curves, allowing their analyze always in a constant range of abscissa $t \in \langle 0; 1 \rangle$. Regardless of the degree of the polynomial and the abscissa x_K of the end point, adequate analyses of the transition curve can be limited to a part that corresponds to the variable $t \in \langle 0; 1 \rangle$. The essence of such change of the variables is illustrated in Fig. 7.9. It shows part of the graph of a polynomial (in this case it comes to the detailed solutions TC_HT-P_GS-1 which are described in Sect. 7.4.1), which corresponds to the classical definition of transition curves. This fragment is marked with a bold line.

7.3.1 First Generalized Solution of Polynomial Transition Curves

The first generalized solution of the polynomial transition curves (GS-1) follows from the function (7.1) by assumption $n = 4$ and assumption that the designed values of y_P, y'_P, y'_K, y''_P and y''_K are known for the points P $(x = 0)$ and K $(x = x_K)$. This solution has a form:

$$y = y_P + y'_P x_K t + \frac{1}{2} y''_P x_K^2 t^2$$

$$+ \left[x_K \left(y'_K - y'_P - y''_P x_K \right) - \frac{1}{3} x_K^2 \left(y'_K - y'_P \right) \right] t^3 \tag{7.5}$$

$$+ \left[-\frac{1}{2} x_K \left(y'_K - y'_P - y''_P x_K \right) + \frac{1}{4} x_K^2 \left(y''_K - y''_P \right) \right] t^4$$

7.3.2   Second Generalized Solution of Polynomial Transition Curves

The second generalized solution of the polynomial transition curves (GS-2) follows from the function (7.1) by assumption $n = 5$ and assumption that the designed values y_P, y_K, y'_P, y'_K, y''_P and y''_K are known for the points P $(x = 0)$ and K $(x = x_K)$. This solution has a form:

$$y = y_P + y'_P x_K t + \frac{1}{2} y''_P x_K^2 t^2$$

$$+ \left[\begin{array}{c} 10 \left(y_K - y_P - y'_P x_K - \frac{1}{2} y''_P x_K^2 \right) - 4 x_K \left(y'_K - y'_P - y''_P x_K \right) \\ + \frac{1}{2} x_K^2 \left(y''_K - y''_P \right) \end{array} \right] t^3$$

$$+ \left[\begin{array}{c} -15 \left(y_K - y_P - y'_P x_K - \frac{1}{2} y''_P x_K^2 \right) + 7 x L_K \left(y'_K - y'_P - y''_P x_K \right) \\ - x_K^2 \left(y''_K - y''_P \right) \end{array} \right] t^4 \tag{7.6}$$

$$+ \left[\begin{array}{c} 6 \left(y_K - y_P - y''_P x_K - \frac{1}{2} y''_P x_K^2 \right) - 3 x_K \left(y'_K - y'_P - y''_P x_K \right) \\ + \frac{1}{2} x_K^2 \left(y''_K - y''_P \right) \end{array} \right] t^5$$

7.3.3   Third Generalized Solution of Polynomial Transition Curves

The third generalized solution of the polynomial transition curves (GS-3) follows from the function (7.1) by assumption $n = 6$ and assumption that the designed values y_P, y_P', y_K', y_P'', y_K'', y_P''' and y_K''' are known for the points P $(x = 0)$ and K $(x = x_K)$. This solution has a form:

$$
\begin{aligned}
y = {}& y_P + y_P' x_K t + \frac{1}{2} y_P'' x_K^2 t^2 + \frac{1}{6} y_P''' x_K^3 t^3 \\
&+ \left[\frac{5}{2} x_K \left(y_K' - y_P' - y_P'' x_K - \frac{1}{2} y_P''' x_K^2 \right) - x_K^2 \left(y_K'' - y_P'' - y_P''' x_K \right) \right. \\
&\qquad\qquad\qquad\qquad\qquad\qquad\qquad\qquad \left. + \frac{1}{8} x_K^3 \left(y_K''' - y_P''' \right) \right] t^4 \\
&+ \left[-3 x_K \left(y_K' - y_P' - y_P'' x_K - \frac{1}{2} y_P''' x_K^2 \right) + \frac{7}{5} x_K^2 \left(y_K'' - y_P'' - y_P''' x_K \right) \right. \\
&\qquad\qquad\qquad\qquad\qquad\qquad\qquad\qquad \left. - \frac{1}{5} x_K^3 \left(y_K''' - y_P''' \right) \right] t^5 \\
&+ \left[x_K \left(y_K' - y_P' - y_P'' x_K - \frac{1}{2} y_P''' x_K^2 \right) - \frac{1}{2} x_K^2 \left(y_K'' - y_P'' - y_P''' x_K \right) \right. \\
&\qquad\qquad\qquad\qquad\qquad\qquad\qquad\qquad \left. + \frac{1}{12} x_K^3 \left(y_K''' - y_P''' \right) \right] t^6
\end{aligned} \tag{7.7}
$$

7.3.4 *Fourth Generalized Solution of Polynomial Transition Curves*

The fourth generalized solution of the polynomial transition curves (GS-2) follows from the function (7.1) by assumption $n = 7$ and assumption that the designed values y_P, y_K, y_P', y_K', y_P'', y_K'', y_P''' and y_K''' are known for the point $P(x = 0)$ and K $(x = x_K)$.

A generalized solution as obtained in this case shows that the result of increasing degree of the polynomial function from $n = 6$ to $n = 7$ is a significant expansion of the curves equation. This is a signal that the potentially increase the degree of the polynomial function to $n \geq 8$ to take into account any additional boundary conditions is practically not justified. For this reason, any implementation of additional boundary conditions for other points than the P and K is only reasonable for the lower degree of the basis polynomial function.

The generalized solution obtained on the basis of the function (7.1) by assumption $n = 7$ has a form:

$$y = y_P + y'_P x_K t + \frac{1}{2} y''_P x_K^2 t^2 + \frac{1}{6} y'''_P x_K^3 t^3$$

$$+ \left[\begin{array}{l} 35\left(y_K - y_P - y'_P x_K - \frac{1}{2} y''_P x_K^2 - \frac{1}{6} y'''_P x_K^3 \right) \\[2mm] -15 x_K \left(y'_K - y'_P - y''_P x_K - \frac{1}{2} y'''_P x_K^2 \right) + \\[2mm] + \frac{5}{2} x_K^2 \left(y''_K - y''_P - y'''_P x_K \right) - \frac{1}{6} x_K^3 \left(y'''_K - y'''_P \right) \end{array} \right] t^4$$

$$+ \left[\begin{array}{l} -84\left(y_K - y_P - y'_P x_K - \frac{1}{2} y''_P x_K^2 - \frac{1}{6} y'''_P x_K^3 \right) \\[2mm] + 39 x_K \left(y'_K - y'_P - y''_P x_K - \frac{1}{2} y'''_P x_K^2 \right) - \\[2mm] -7 x_K^2 \left(y''_K - y''_P - y'''_P x_K \right) + \frac{1}{2} x_K^3 \left(y'''_K - y'''_P \right) \end{array} \right] t^5 \qquad (7.8)$$

$$+ \left[\begin{array}{l} 70\left(y_K - y_P - y'_P x_K - \frac{1}{2} y''_P x_K^2 - \frac{1}{6} y'''_P x_K^3 \right) \\[2mm] -34 x_K \left(y'_K - y'_P - y''_P x_K - \frac{1}{2} y'''_P x_K^2 \right) + \\[2mm] + \frac{13}{2} x_K^2 \left(y''_K - y''_P - y'''_P x_K \right) - \frac{1}{2} x_K^3 \left(y'''_K - y'''_P \right) \end{array} \right] t^6$$

$$+ \left[\begin{array}{l} -20\left(y_K - y_P - y'_P x_K - \frac{1}{2} y''_P x_K^2 - \frac{1}{6} y'''_P x_K^3 \right) \\[2mm] + 10 x_K \left(y'_K - y'_P - y''_P x_K - \frac{1}{2} y'''_P x_K^2 \right) - \\[2mm] -2 x_K^2 \left(y''_K - y''_P - y'''_P x_K \right) + \frac{1}{6} x_K^3 \left(y'''_K - y'''_P \right) \end{array} \right] t^7$$

7.4 Different Solutions of Polynomial Transition Curves

Generalized solutions as described in Sect. 7.3, can be used to define the respective polynomial equations of transition curves that belong to one category of the curves listed at the beginning of this chapter. The form of these equations depend on the degree of the basis polynomial function and acceptable assumptions with respect to y_P, y_K, y'_P, y'_K, y''_P, y''_K, y'''_P and y'''_K for the points P $(x = 0)$ and K $(x = x_K)$. These assumptions should be regarded as the boundary conditions that define the geometry of the particular transition curves.

The possible boundary conditions are shown in Table 7.1. This statement, of course, does not exhaust the possibility of assuming the following conditions, among others, for other points than the P and the K. However, it should be noted that the assumption of additional conditions requires a corresponding increase in the degree of the polynomial function (7.1), which is the basis for the establishment for identification of the curve equation.

Accordingly to the Figs. 7.1, 7.2, 7.3, 7.4, 7.5, 7.6, 7.7 and 7.8, the following notations under the conditions given in Table 7.1 were assumed:

R_P designed radius of curvature at the point P
R_K designed radius of curvature at the point K
$\tan u_P$ designed inclination of tangent at the point P
$\tan u_K$ designed inclination of tangent at the point K

A sign "−" or "+" is used to define the convexity or concavity of the arc at the points P and K.

In the case of any specific solutions, it is necessary to make an analysis, which is to establish the design conditions that ensure qualifying individual curves to a specific category. This analysis involves the use of special procedures in order to evaluate graphs of derivatives y', y'' and y''' and formulate appropriate design conditions. All conditions have been given on the basis of work of Kobryń (2009). A method for determining the design conditions will be illustrated in relation to one of specific solutions, which are described later in this section.

Tables with the characteristics of the individual curves, which are presented later in this chapter, include the design conditions that result from this analysis with respect to these curves. In order to uniquely identify of these curves, the appropriate identification codes there have been used. The first part of the code is an

Table 7.1 Possible boundary conditions for defining of polynomial transition curves

Point	Form of condition
P	$y(x = 0) = 0$
	$y'(x = 0) = 0$
	$y'(x = 0) = \tan u_P$
	$y''(x = 0) = 0$
	$y''(x = 0) = \pm \frac{1}{R_P}\left(1 + \tan^2 u_P\right)^{3/2}$
	$y'''(x = 0) = 0$
K	$y(x = x_K) = 0$
	$y(x = x_K) = y_K$
	$y'(x = x_K) = 0$
	$y'(x = x_K) = \tan u_K$
	$y''(x = x_K) = 0$
	$y''(x = x_K) = -\frac{1}{R_K}$
	$y''(x = x_K) = \pm \frac{1}{R_K}\left(1 + \tan^2 u_K\right)^{3/2}$
	$y'''(x = x_K) = 0$

abbreviation of the appropriate curves category, while the second part—the designation of generalized solutions, from which were received the specific solution of transition curves.

7.4.1   Polynomial Transition Curves Based on the First Generalized Solution

On the basis of generalized solution GS-1, following specific solutions can be defined:

- transition curves with a horizontal tangent at the point P (code TC_HT-P_GS-1)
- transition curves with a horizontal tangent at the point K (code TC_HT-K_GS-1)
- general transition curves (code GTC_GS-1)
- universal and oval transition curves (code UOTC_GS-1)

Characteristics of these solutions were presented in Tables 7.2, 7.3, 7.4 and 7.5.

Apart from the equations of the curves, these tables also contain the appropriate design conditions. These conditions are necessary to obtain a distribution of curvature that is appropriate to the category of curves.

A method of determining the design conditions will be presented on the example of transition curves with a horizontal tangent at the point P (identification code TC_HT-P_GS-1). For these curves, the curvature should increase from zero at the point P to value $1/R_K$ at the point K. To achieve this, the derivative y' should be increasing and convex in the entire range of values $t \in \langle 0; 1 \rangle$. It follows that the function should not have extreme points and inflection points. Therefore, it is necessary to analyze the derivatives y'' and y'''.

Derivative y'' has the form:

$$y'' = \frac{\tan u_K}{x_K} \cdot N_2'' + \frac{1}{R_K}\left(1 + \tan^2 u_K\right)^{3/2} N_4'' \tag{7.9}$$

Table 7.2 Characteristics of the curves TC_HT-P_GS-1

Curves category	Transition curves (with a horizontal tangent at point P)
Generalized solution	GS-1
Identification code	TC_HT-P_GS-1
Equation	$y = x_K \tan u_K \cdot N_2 + \frac{x_K^2}{R_K}\left(1 + \tan^2 u_K\right)^{3/2} N_4$ $N_2 = t^3 - \frac{1}{2}t^4$ $N_4 = -\frac{1}{3}t^3 + \frac{1}{4}t^4$
Design conditions	$D = \dfrac{R_K \tan u_K}{x_K\left(1 + \tan^2 u_K\right)^{3/2}}$ $D \in \langle 1/3; 2/3 \rangle$ $\tan u_K = \sqrt{\dfrac{4D - 6D^2}{3 - 4D + 6D^2}}$

Table 7.3 Characteristics of the curves TC_HT-K_GS-1

Curves category	Transition curves (with a horizontal tangent at point K)
Identification code	TC_HT-K_GS-1
Generalized solution	GS-1
Equation	$y = x_K \tan u_P \cdot N_1 - \frac{x_K^2}{R_K} N_4$ $N_1 = t - t^3 + \frac{1}{2}t^4$ $N_4 = -\frac{1}{3}t^3 + \frac{1}{4}t^4$
Design conditions	$E = \frac{R_K \tan u_P}{x_K}$ $E \in \langle 1/3; 2/3 \rangle$

Table 7.4 Characteristics of the curves GTC_GS-1

Curves category	General transition curves
Generalized solution	GS-1
Identification code	GTC_GS-1
Equation	$y = x_K \tan u_P \cdot N_1 + x_K \tan u_K \cdot N_2$ $N_1 = t - t^3 + \frac{1}{2}t^4$ $N_2 = t^3 - \frac{1}{2}t^4$
Design conditions	Without restrictions

Table 7.5 Characteristics of the curves UOTC_GS-1

Curves category	Universal (and oval) transition curves
Generalized solution	GS-1
Identification code	UOTC_GS-1
Equation	$y = x_K \tan u_P \cdot N_1 + x_K \tan u_K \cdot N_2$ $+ \frac{x_K^2}{R_P}\left(1 + \tan^2 u_P\right)^{3/2} n_P N_3 + \frac{x_K^2}{R_K}\left(1 + \tan^2 u_K\right)^{3/2} n_K N_4$ $N_1 = t - t^3 + \frac{1}{2}t^4$ $N_2 = t^3 - \frac{1}{2}t^4$ $N_3 = \frac{1}{2}t^2 - \frac{2}{3}t^3 + \frac{1}{4}t^4$ $N_4 = -\frac{1}{3}t^3 + \frac{1}{4}t^4$ $\bullet$ $y_P'' < 0 \Rightarrow n_P = -1$ $y_P'' > 0 \Rightarrow n_P = +1$ $\bullet$ $y_K'' < 0 \Rightarrow n_K = -1$ $y_K'' > 0 \Rightarrow n_K = +1$

(continued)

Table 7.5 (continued)

Curves category	Universal (and oval) transition curves
Design conditions	Universal transition curves:

$$\tan u_P - \tan u_K = n_P N_{3/1}^{(II)} \frac{x_K}{R_P} \left(1 + \tan^2 u_P\right)^{3/2}$$

$$+ n_K N_{4/1}^{(II)} \frac{x_K}{R_K} \left(1 + \tan^2 u_K\right)^{3/2}$$

$$n_P \cdot n_K = -1$$

$$N_{3/1}^{(II)} \in \langle -1/3; \infty)$$

$$N_{4/1}^{(II)} \in \langle -1/3; \infty)$$

Oval transition curves:

$$\tan u_P - n_{2/1}^{(II)} \tan u_K = n_P n_{3/1}^{(II)} \frac{x_K}{R_P} \left(1 + \tan^2 u_P\right)^{3/2}$$

$$+ n_K n_{4/1}^{(II)} \frac{x_K}{R_K} \left(1 + \tan^2 u_K\right)^{3/2}$$

$$n_P \cdot n_K = 1$$

$$n_{2/1}^{(II)} \neq 1$$

$$n_{3/1}^{(II)} \in (-\infty; -1/3)$$

$$n_{4/1}^{(II)} \in (-\infty; -1/3)$$

wherein

$$N_2'' = 6t - 6t^2,$$
$$N_4'' = -2t + 3t^2$$

Derivative y''' has the form:

$$y''' = \frac{\tan u_K}{x_K^2} \cdot N_2''' + \frac{1}{x_K R_K} \left(1 + \tan^2 u_K\right)^{3/2} N_4''' \tag{7.10}$$

wherein

$$N_2''' = 6 - 12t,$$
$$N_4''' = -2 + 6t$$

From condition $y'' = 0$ it follows:

$$\frac{R_K \tan u_K}{x_K \left(1 + \tan^2 u_K\right)^{3/2}} = -\frac{N_4''}{N_2''} = N_{4/2}'' \tag{7.11}$$

Graph of the function $N_{4/2}''$ is shown in Fig. 7.10. The graph of the derivative y' has no inflection points if $y'' \neq 0$. From Eq. (7.11) and Fig. 7.10 it follows that a condition for this is fulfillment of an inequality:

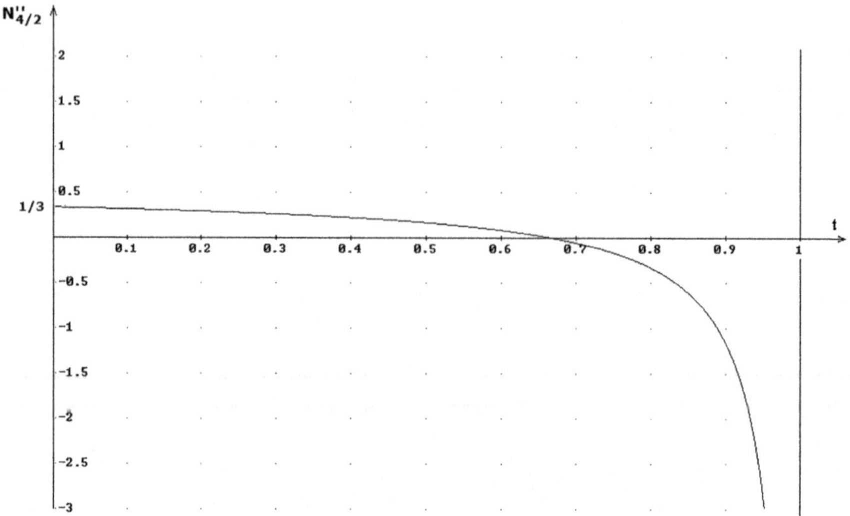

Fig. 7.10 Graph of the function $N''_{4/2}$

$$\frac{1}{3} \leq \frac{R_K \tan u_K}{x_K(1+\tan^2 u_K)^{3/2}} \tag{7.12}$$

From condition $y''' = 0$ it follows:

$$\frac{R_K \tan u_K}{x_K(1+\tan^2 u_K)^{3/2}} = -\frac{N'''_4}{N'''_2} = N'''_{4/2} \tag{7.13}$$

Graph of the function $N'''_{4/2}$ is shown in Fig. 7.11. The derivative y' is increasing if $y'' > 0$. Therefore, the derivative y' should not have inflection points in the interval $t \in \langle 0; 1 \rangle$. Hence it follows $y''' \neq 0$. From Eq. (7.13) and Fig. 7.11 result that derivative y' Has no inflection points in interval $t \in \langle 0; 1 \rangle$, if:

$$\frac{1}{3} \leq \frac{R_K \tan u_K}{x_K(1+\tan^2 u_K)^{3/2}} \leq \frac{2}{3} \tag{7.14}$$

However, from graph of the function $N''_{4/2}$ (Fig. 7.10) it follows that $y'' > 0$, if the condition (7.11) is fulfilled. Finally, it follows a total condition in the form $D \in \langle 1/3; 2/3 \rangle$, whereby:

$$D = \frac{R_K \tan u_K}{x_K(1+\tan^2 u_K)^{3/2}} \tag{7.15}$$

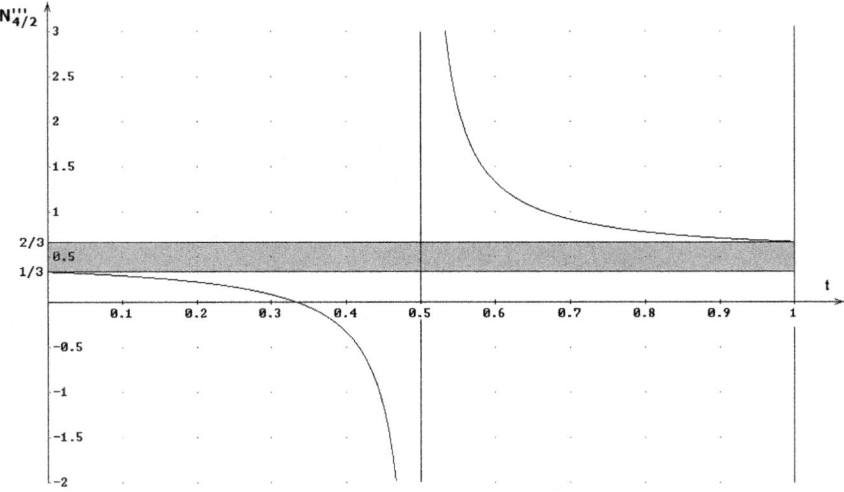

Fig. 7.11 Graph of the function $N_{4/2}^{'''}$

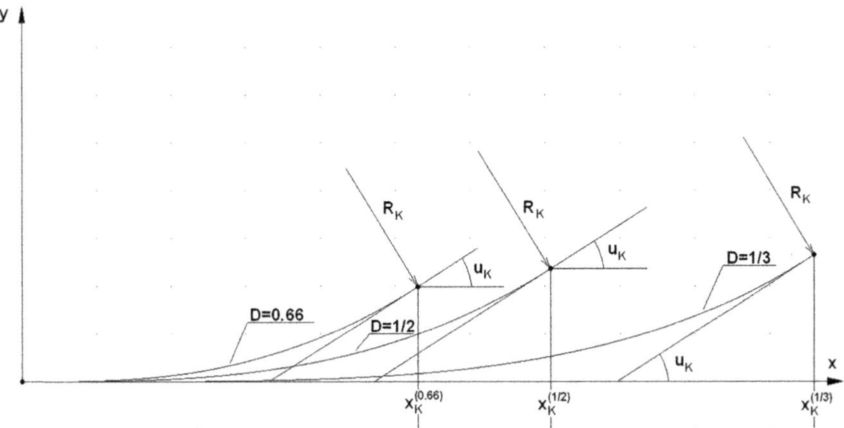

Fig. 7.12 Exemplary graphs of curves TC_HT-P_GS-1 for selected values D

It can be added that a selection of the value $D \in \langle 1/3; 2/3 \rangle$ allows a freedom of shaping the curves geometry, as is illustrated in Fig. 7.12. It show a graphs of the curves for selected values D.

In a similar manner as discussed above for the curves TC_HT-P_GS-1, appropriate design conditions may be determined for other transition curves. Appropriate analyzes were presented in the work Kobryń (2009), and the resulting conditions are given in the following tables in Sect. 7.4.

7.4.2 Polynomial Transition Curves Based on the Second Generalized Solution

On the basis of generalized solution GS-2, can be defined specific solutions:

- quasi-transition curves with a horizontal tangent at point P (code QTC_HT-P_GS-2)
- quasi-transition curves with a horizontal tangent at point K (code QTC_HT-K_GS-2)
- general and S-shaped transition curves (code GSSTC_GS-2)
- universal and oval transition curves (code UOTC_GS-2)

The characteristics of these solutions is shown in the Tables 7.6, 7.7, 7.8 and 7.9.

7.4.3 Polynomial Transition Curves Based on the Third Generalized Solution

On the basis of generalized solution GS-3, can be defined specific solutions:

- transition curves with a horizontal tangent at point K (code TC_HT-K_GS-3)
- quasi-transition curves with a horizontal tangent at the P (QTC_HT-P_GS-3)
- general transition curves (code GTC_GS-3)
- universal and oval transition curves (code UOTC_GS-3)

The characteristics of these solutions is shown in the Tables 7.10, 7.11, 7.12 and 7.13.

Table 7.6 Characteristics of the curves QTC_HT-P_GS-2

Curves category	Quasi-transition curves (with a horizontal tangent at point P)
Generalized solution	GS-2
Identification code	QTC_HT-P_GS-2
Equation	$y = x_K \tan \alpha \cdot M_0 + x_K \tan u_K \cdot M_2 + \frac{x_K^2}{R_K}\left(1+\tan^2 u_K\right)^{3/2} M_4$ $M_0 = 10t^3 - 15t^4 + 6t^5$ $M_2 = -4t^3 + 7t^4 - 3t^5$ $M_4 = \frac{1}{2}t^3 - t^4 + \frac{1}{2}t^5$ $\tan \alpha = y_K/x_K$
Design conditions	No conditions; selection of curve parameters on the basis of graph of function: $\tan \alpha = -\frac{1}{M_0''}\left[M_2'' \tan u_K + M_4'' \frac{x_K}{R_K}\left(1+\tan^2 u_K\right)^{3/2}\right]$ $\tan \alpha = -\frac{1}{M_0'''}\left[M_2''' \tan u_K + M_4''' \frac{x_K}{R_K}\left(1+\tan^2 u_K\right)^{3/2}\right]$

Table 7.7 Characteristics of the curves QTC_HT-K_GS-2

Curves category	Quasi-transition curves (with a horizontal tangent at point K)
Generalized solution	GS-2
Identification code	QTC_HT-K_GS-2
Equation	$y = x_K \tan\alpha \cdot M_0 + x_K \tan u_P \cdot M_1 - \frac{x_K^2}{R_K} M_4$ $M_0 = 10t^3 - 15t^4 + 6t^5$ $M_1 = t - 6t^3 + 8t^4 - 3t^5$ $M_4 = \frac{1}{2}t^3 - t^4 + \frac{1}{2}t^5$ $\tan\alpha = y_K/x_K$
Design conditions	No conditions; selection of curve parameters using graph of function: $\tan\alpha = \frac{1}{M_0''}\left[M_4'' \frac{x_K}{R_K} - M_1'' \tan u_P\right]$ $\tan\alpha = \frac{1}{M_0'''}\left[M_4''' \frac{x_K}{R_K} - M_1''' \tan u_P\right]$

Table 7.8 Characteristics of the curves GSSTC_GS-2

Curves category	General and S-shaped transition curves
Generalized solution	GS-2
Identification code	GSSTC_GS-2
Equation	$y = x_K \tan\alpha \cdot M_0 + x_K \tan u_P \cdot M_1 + x_K \tan u_K \cdot M_2$ $M_0 = 10t^3 - 15t^4 + 6t^5$ $M_1 = t - 6t^3 + 8t^4 - 3t^5$ $M_2 = -4t^3 + 7t^4 - 3t^5$
Design conditions	General transition curves: $\tan u_P \cdot \tan u_K < 0$ $\tan\alpha = M_{1/0}^{(II)}\tan u_P + M_{2/0}^{(II)}\tan u_K$ S-shaped transition curves: $\tan u_P \cdot \tan u_K > 0$ where: $M_{1/0}^{(II)} \in \langle 2/5; 3/5\rangle; M_{2/0}^{(II)} \in \langle 2/5; 3/5\rangle$

7.4.4   Polynomial Transition Curves Based on the Fourth Generalized Solution

On the basis of generalized solution GS-4, can be defined specific solutions:

- quasi-transition curves with a horizontal tangent at point P (code QTC_HT-P_GS-4)
- quasi-transition curves with a horizontal tangent at point K (code QTC_HT-K_GS-4)

Table 7.9 Characteristics of the curves UOTC_GS-2

Curves category	Universal (and oval) transition curves
Generalized solution	GS-2
Identification code	UOTC_GS-2
Equation	$y = x_K \tan\alpha \cdot M_0 + x_K \operatorname{tg} u_P \cdot M_1 + x_K \tan u_K \cdot M_2$ $\quad + \dfrac{x_K^2}{R_P}\left(1+\tan^2 u_P\right)^{3/2} m_P M_3 + \dfrac{x_K^2}{R_K}\left(1+\tan^2 u_K\right)^{3/2} m_K M_4$ $M_0 = 10t^3 - 15t^4 + 6t^5$ $M_1 = t - 6t^3 + 8t^4 - 3t^5$ $M_2 = -4t^3 + 7t^4 - 3t^5$ $M_3 = \dfrac{1}{2}t^2 - \dfrac{3}{2}t^3 + \dfrac{3}{5}t^4 - \dfrac{1}{2}t^5$ $M_4 = \dfrac{1}{2}t^3 - t^4 + \dfrac{1}{2}t^5$ • $y_P'' < 0 \Rightarrow m_P = -1$ $\quad y_P'' > 0 \Rightarrow m_P = +1$ • $y_K'' < 0 \Rightarrow m_K = -1$ $\quad y_K'' > 0 \Rightarrow m_K = +1$
Design conditions	Selection of curve parameters using graph of function: $$V = -\frac{m_K M_4''\left(1+\tan^2 u_K\right)^{3/2}}{U\left(M_0'' \tan\alpha + M_1'' \tan u_P + M_2'' \tan u_K\right) + m_P M_3''\left(1+\tan^2 u_P\right)^{3/2}}$$ $$V = -\frac{m_K M_4'''\left(1+\tan^2 u_K\right)^{3/2}}{U\left(M_0''' \tan\alpha + M_1''' \tan u_P + M_2''' \tan u_K\right) + m_P M_3'''\left(1+\tan^2 u_P\right)^{3/2}}$$ or $$U = -\frac{m_K M_4''\left(1+\tan^2 u_K\right)^{3/2} + V m_P M_3''\left(1+\tan^2 u_P\right)^{3/2}}{V\left(M_0'' \tan\alpha + M_1'' \tan u_P + M_2'' \tan u_K\right)}$$ $$U = -\frac{m_K M_4'''\left(1+\tan^2 u_K\right)^{3/2} + V m_P M_3'''\left(1+\tan^2 u_P\right)^{3/2}}{V\left(M_0''' \tan\alpha + M_1''' \tan u_P + M_2''' \tan u_K\right)}$$
	$U = R_P/x_K$ $V = R_K/R_P$ $m_P \cdot m_K = -1$ (universal) $m_P \cdot m_K = 1$ (oval)

Table 7.10 Characteristics of the curves TC_HT-K_GS-3

Curves category	Transition curves (with a horizontal tangent at point K)
Generalized solution	GS-3
Identification code	TC_HT-K_GS-3
Equation	$y = x_K \tan u_P \cdot F_1 - \dfrac{x_K^2}{R_P} F_4$ $F_1 = t - \frac{5}{2}t^4 + 3t^5 - t^6$ $F_4 = -t^4 + \frac{7}{5}t^5 - \frac{1}{2}t^6$
Design conditions	$E = \dfrac{R_K \tan u_P}{x_K}$ $E \in \langle 4/10;\, 6/10\rangle$

Table 7.11 Characteristics of the curves QTC_HT-P_GS-3

Curves category	Quasi-transition curves (with a horizontal tangent at point P)
Generalized solution	GS-3
Identification code	QTC_HT-P_GS-3
Equation	$y = x_K \tan u_K \cdot F_2 + \frac{x_K^2}{R_K}(1 + \tan^2 u_K)^{3/2} F_4$ $F_2 = \frac{5}{2} t^4 - 3t^5 + t^6$ $F_4 = -t^4 + \frac{7}{5} t^5 - \frac{1}{2} t^6$
Design conditions	$D = \dfrac{R_K \tan u_K}{x_K (1 + \tan^2 u_K)^{3/2}}$ $D \geq 4/10$

Table 7.12 Characteristics of the curves GTC_GS-3

Curves category	General transition curves
Generalized solution	GS-3
Identification code	GTC_GS-3
Equation	$y = x_K \tan u_P \cdot F_1 + x_K \tan u_K \cdot F_2$ $F_1 = t - \frac{5}{2} t^4 + 3t^5 - t^6$ $F_2 = \frac{5}{2} t^4 - 3t^5 + t^6$
Design conditions	Without restrictions

- general transition curves (code GSSTC_GS-4)
- universal and oval transition curves (code UOTC_GS-4)

The characteristics of these solutions is shown in the Tables 7.14, 7.15, 7.16 and 7.17.

7.5   Selection of Design Parameters for Universal and Oval Transition Curves

In general, it is possible that the determined equation of the curve, which can be potentially classified to a specific category of curves, will include undesirable extremes of curvature in the range of the variable $t \in \langle 0; 1 \rangle$. Therefore, the design conditions are given in the tables in Sect. 7.4, which ensure the achievement of the correct distribution of curvature. However, in the case of the universal and oval transition curves additional procedures are necessary to obtain a curve graph that corresponds to the category of the curves. This is explained below.

Table 7.13 Characteristics of the curves UOTC_GS-3

Curves category	Universal and oval transition curves
Generalized solution	GS-3
Identification code	UOTC_GS-3
Equation	$y = x_K \tan u_P \cdot F_1 + x_K \tan u_K \cdot F_2$ $+ \dfrac{x_K^2}{R_P}\left(1 + \tan^2 u_P\right)^{3/2} f_P F_3 + \dfrac{x_K^2}{R_K}\left(1 + \tan^2 u_K\right)^{3/2} f_K F_4$ $F_1 = t - \dfrac{5}{2} t^4 + 3t^5 - t^6$ $F_2 = \dfrac{5}{2} t^4 - 3t^5 + t^6$ $F_3 = \dfrac{1}{2} t^2 - \dfrac{3}{2} t^4 + \dfrac{8}{5} t^5 - \dfrac{1}{2} t^6$ $F_4 = -t^4 + \dfrac{7}{5} t^5 - \dfrac{1}{2} t^6$ • $\begin{aligned} y_P'' &< 0 \Rightarrow f_P = -1 \\ y_P'' &> 0 \Rightarrow f_P = +1 \end{aligned}$ • $\begin{aligned} y_K'' &< 0 \Rightarrow f_K = -1 \\ y_K'' &> 0 \Rightarrow f_K = +1 \end{aligned}$
Design conditions	Universal transition curves: $\tan u_P - \tan u_K$ $= f_P F_{3/1}^{(II)} \dfrac{x_K}{R_P}\left(1 + \tan^2 u_P\right)^{3/2} + f_K F_{4/1}^{(II)} \dfrac{x_K}{R_K}\left(1 + \tan^2 u_K\right)^{3/2}$ $f_P \cdot f_K = -1 \ \dots \ F_{3/1}^{(II)} \in \langle -2/5;\ \infty) \ \dots \ F_{4/1}^{(II)} \in \langle -2/5;\ \infty)$ Oval transition curves: $\tan u_P - f_{2/1}^{(II)} \tan u_K$ $= f_P f_{3/1}^{(II)} \dfrac{x_K}{R_P}\left(1 + \tan^2 u_P\right)^{3/2} + f_K f_{4/1}^{(II)} \dfrac{x_K}{R_K}\left(1 + \tan^2 u_K\right)^{3/2}$ $f_P \cdot f_K = 1 \quad f_{2/1}^{(II)} \neq 1 \quad f_{3/1}^{(II)} \in (-\infty;\ -2/5) \quad f_{4/1}^{(II)} \in (-\infty;\ -2/5)$

Table 7.14 Characteristics of the curves QTC_HT-P_GS-3

Curves category	Quasi-transition curves (with a horizontal tangent at point P)
Generalized solution	GS-4
Identification code	QTC_HT-P_GS-3
Equation	$y = x_K \tan \alpha \cdot G_0 + x_K \tan u_K \cdot G_2 + \dfrac{x_K^2}{R_K}\left(1 + \tan^2 u_K\right)^{3/2} G_4$ $G_0 = 35t^4 - 84t^5 + 70t^6 - 20t^7$ $G_2 = -15t^4 + 39t^5 - 34t^6 + 10t^7$ $G_4 = \tfrac{5}{2} t^4 - 7t^5 + \tfrac{13}{2} t^6 - 2t^7$ $\tan \alpha = y_K / x_K$
Design conditions	No conditions; selection of curve parameters using graph of function: $\tan \alpha = -\dfrac{1}{G_0''}\left[G_2'' \tan u_K + G_4'' \dfrac{x_K}{R_K}\left(1 + \tan^2 u_K\right)^{3/2} \right]$ $\tan \alpha = -\dfrac{1}{G_0'''}\left[G_2''' \tan u_K + G_4''' \dfrac{x_K}{R_K}\left(1 + \tan^2 u_K\right)^{3/2} \right]$

Table 7.15 Characteristics of the curves QTC_HT-K_GS-4

Curves category	Quasi-transition curves (with a horizontal tangent at point K)
Generalized solution	GS-4
Identification code	QTC_HT-K_GS-4
Equation	$y = x_K \tan\alpha \cdot G_0 + x_K \tan u_P \cdot G_1 - \frac{x_K^2}{R_K} G_4$ $G_0 = 35t^4 - 84t^5 + 70t^6 - 20t^7$ $G_1 = t - 20t^4 + 45t^5 - 36t^6 + 10t^7$ $G_4 = \frac{5}{2}t^4 - 7t^5 + \frac{13}{2}t^6 - 2t^7$ $\tan\alpha = y_K/x_K$
Design conditions	No conditions; selection of curve parameters on the basis of graph of function: $\tan\alpha = \frac{1}{G_0''}\left[G_4'' \frac{x_K}{R_K} - G_1'' \tan u_P\right]$ $\tan\alpha = \frac{1}{G_0'''}\left[G_4''' \frac{x_K}{R_K} - G_1''' \tan u_P\right]$

Table 7.16 Characteristics of the curves GSSTC_GS-4

Curves category	General and S-shaped transition curves
Generalized solution	GS-4
Identification code	GSSTC_GS-4
Equation	$y = x_K \tan\alpha \cdot G_0 + x_K \tan u_P \cdot G_1 + x_K \tan u_K \cdot G_2$ $G_0 = 35t^4 - 84t^5 + 70t^6 - 20t^7$ $G_1 = t - 20t^4 + 45t^5 - 36t^6 + 10t^7$ $G_2 = t - 15t^4 + 39t^5 - 34t^6 + 10t^7$
Design conditions	General transition curves: $\tan u_P \cdot \tan u_K < 0$ $\tan\alpha = G_{1/0}^{(II)}\tan u_P + G_{2/0}^{(II)}\tan u_K$ S-shaped transition curves: $\tan u_P \cdot \tan u_K > 0$ where: $G_{1/0}^{(II)} \in \langle 3/7; 4/7\rangle$ $G_{2/0}^{(II)} \in <3/7; 4/7>$

For example, in the case of the curves with the code UOTC_GS-2 we assume that: $\tan u_P = 3$, $\tan u_K = 2$, $\tan\alpha = 0.5$, $U = 2$. Based on these assumptions, from the graphs of function V (Fig. 7.13) can be read for example $V = 0.5$. The read value V provides a single point of inflection and an absence extremes of derivative y'' on each of the two sections, to which the curve is divided by an inflection point Q (Fig. 7.7). Assuming that $R_P = 2000$ m, it follows $x_K = 1000$ m and $R_K = 1000$ m.

Table 7.17 Characteristics of the curves UOTC_GS-4

Curves category	Universal and oval transition curves
Generalized solution	GS-4
Identification code	UOTC_GS-4
Equation	$y = x_K \tan\alpha \cdot G_0 + x_K \tan u_P \cdot G_1 + x_K \tan u_K \cdot G_2 +$ $$+ \frac{x_K^2}{R_P}\left(1 + \tan^2 u_P\right)^{3/2} g_P G_3 + \frac{x_K^2}{R_K}\left(1 + \tan^2 u_K\right)^{3/2} g_K G_4$$ $$G_0 = 35t^4 - 84t^5 + 70t^6 - 20t^7$$ $$G_1 = t - 20t^4 + 45t^5 - 36t^6 + 10t^7$$ $$G_2 = -15t^4 + 39t^5 - 34t^6 + 10t^7$$ $$G_3 = \frac{1}{2}t^2 - 5t^4 + 10t^5 - \frac{15}{2}t^6 + 2t^7$$ $$G_4 = \frac{5}{2}t^4 - 7t^5 + \frac{13}{2}t^6 - 2t^7$$ $$\bullet \quad \begin{aligned} y_P'' < 0 &\Rightarrow g_P = -1 \\ y_P'' > 0 &\Rightarrow g_P = +1 \end{aligned}$$ $$\bullet \quad \begin{aligned} y_K'' < 0 &\Rightarrow g_K = -1 \\ y_K'' > 0 &\Rightarrow g_K = +1 \end{aligned}$$
Design conditions	$$V'' = -\frac{g_K G_4''\left(1 + \tan^2 u_K\right)^{3/2}}{U\left(G_0'' \tan\alpha + G_1'' \tan u_P + G_2'' \tan u_K\right) + g_P G_3''\left(1 + \tan^2 u_P\right)^{3/2}}$$ $$V''' = -\frac{g_K G_4'''\left(1 + \tan^2 u_K\right)^{3/2}}{U\left(G_0''' \tan\alpha + G_1''' \tan u_P + G_2''' \tan u_K\right) + g_P G_3'''\left(1 + \tan^2 u_P\right)^{3/2}}$$ or $$U'' = -\frac{g_K G_4''\left(1 + \tan^2 u_K\right)^{3/2} + V g_P G_3''\left(1 + \tan^2 u_P\right)^{3/2}}{V\left(G_0'' \tan\alpha + G_1'' \tan u_P + G_2'' \tan u_K\right)}$$ $$U''' = -\frac{g_K G_4'''\left(1 + \tan^2 u_K\right)^{3/2} + V g_P G_3'''\left(1 + \tan^2 u_P\right)^{3/2}}{V\left(G_0''' \tan\alpha + G_1''' \tan u_P + G_2''' \tan u_K\right)}$$
	$U = R_P/x_K$ $V = R_K/R_P$ $g_P \cdot g_K = -1$ (universal) $g_P \cdot g_K = 1$ (oval)

Graph of the curve defined by this set of parameters, as well as a graph of its curvature, is presented in Fig. 7.14. A value $t_Q = 0.54002$ corresponds to a point of inflection Q. Within a curve exist two local extrema of curvature that position is given by the values: $t_1 = 0.30628$ and $t_2 = 0.77399$. The values of the curvature radii at the extreme points are $R_1 = 257.62$ m and $R_2 = 259.82$ m.

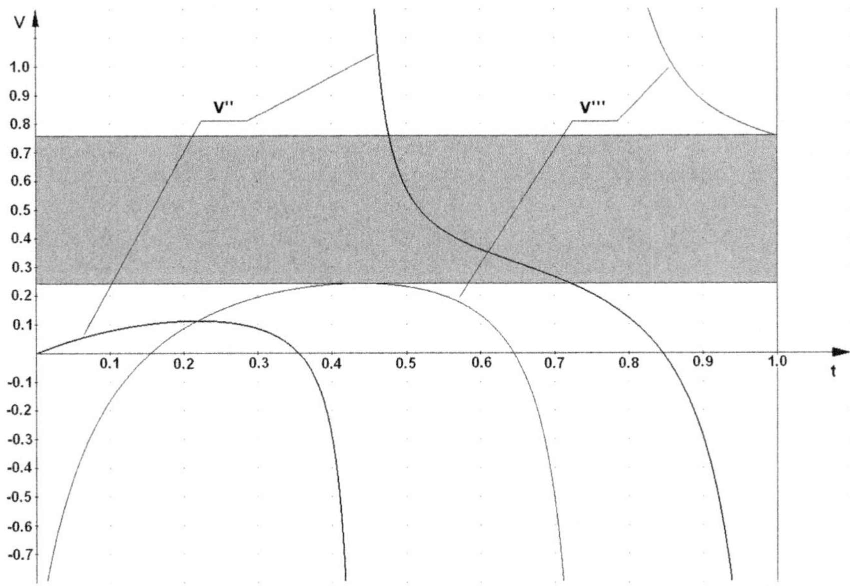

Fig. 7.13 Graphs of functions V'' and V''' for curves UOTC_GS-2

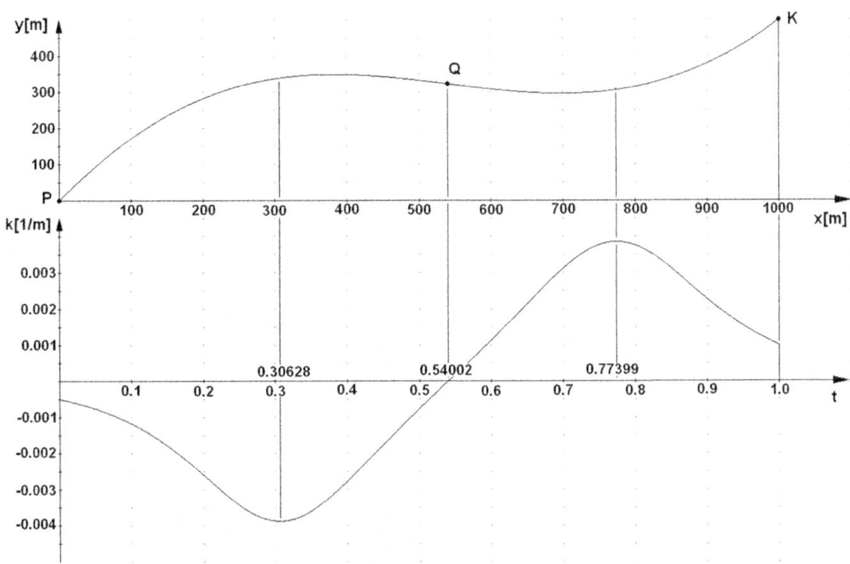

Fig. 7.14 Graph of the curve UOTC_GS-2 and her curvature for a set of parameters defined by the functions V'' and V'''

References

Grabowski RJ (1984) Gładkie przejścia krzywoliniowe w drogach kołowych i kolejowych. Zeszyty Naukowe AGH, Geodezja nr 82, Kraków (in Polish)

Kobryń A (2009) Wielomianowe kształtowanie krzywych przejściowych. Wydawnictwa Politechniki Białostockiej, Rozprawy Naukowe nr 167, Białystok (in Polish)

Chapter 8
Sample Applications of Transition Curves in Horizontal Alignment

Transition curves will be used in horizontal alignment as traditionally understood geometrical systems as well as independent geometric elements that allow you to determine a whole curvilinear transition between two straight lines. Appropriate solutions of this problem will be presented in this section.

8.1 General Remarks

Two types of geometric systems occur generally in traditional applications of transition curves in the design of horizontal curves:

- 1st transition curve—circular arc—2nd transition curve
- 1st transition curve—2nd transition curve

Many works in the literature describe the designing of this type geometric systems using the spiral curve (Brockenbrough 2009; Easa 2003; Lamm et al. 1999; Lipiński 1993; Lorenz 1971; Meyer and Gibson 1980; Rogers 2008; Wolhuter 2015). Calculating each geometric elements was carried out in accordance with Fig. 8.1.

The following designations were assumed:

P   the point of change from tangent to transition curve (start point of the transition curve)
K   the point of change from transition curve do circular curve
S   the centre of the circular curve
W   the point where the normal at the K and tangent at the P intersect
R_K the radius of the circular curie (radius at the K)
$\hat{u}$ the angle between the tangent at the K and the tangent at the P
X_K the abscissa of the K (tangent distance from P to the K)
Y_K the ordinate of the K (tangent offset of the K)
X_S the abscissa of the S

© Springer International Publishing AG 2017

A. Kobryń, *Transition Curves for Highway Geometric Design*, Springer Tracts on Transportation and Traffic 14, DOI 10.1007/978-3-319-53727-6_8

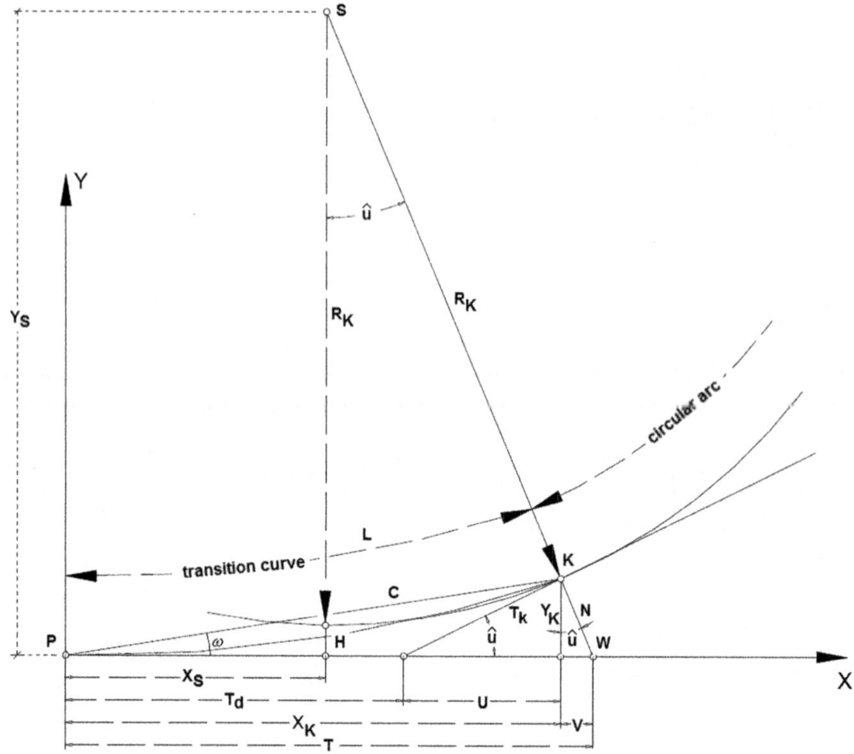

Fig. 8.1 Transition curve as the element connecting straight line and circular arc

Y_S the ordinate of the S

H the ordinate of the offsetted K (ordinate of the offsetted point of curvature, where the circular curve begins)

T the tangent distance from the P to the W (main tangent)

T_d the distance on the tangent from the P to the intersection with a tangent through the K (the longer transition curve tangent)

T_k the distance on the tangent through the K from the K to the intersection with the tangent through the P (the shorter transition curve tangent)

N the distance from the K to the W (normal to the tangent at the K)

U the distance from the K to the intersection of tangents at the P and the K (subtangent)

V the distance from the K to the intersection of tangents at the K and the P (subnormal)

C the chord from the P to the K

ω the angle between the chord from the P and the tangent at the P

In the case of the spiral curve, the following, commonly known formulas are used to calculate mentioned geometrical elements:

$$X_S = X_K - R_K \sin \hat{u} \tag{8.1}$$

$$H = Y_K - R_K(1 - \cos \hat{u}) \tag{8.2}$$

$$Y_S = R_K + H = Y_K + R_K \cos \hat{u} \tag{8.3}$$

$$T = X_K + Y_K \tan \hat{u} \tag{8.4}$$

$$T_d = X_K - Y_K \frac{1}{\tan \hat{u}} \tag{8.5}$$

$$T_k = Y_K \frac{1}{\sin \hat{u}} \tag{8.6}$$

$$N = Y_K \frac{1}{\cos \hat{u}} \tag{8.7}$$

$$U = Y_K \frac{1}{\tan \hat{u}} \tag{8.8}$$

$$V = Y_K \tan \hat{u} \tag{8.9}$$

$$C = \sqrt{X_K^2 + Y_K^2} \tag{8.10}$$

$$\omega = \arctan \frac{Y_K}{X_K} \tag{8.11}$$

where X_K and Y_K are coordinates of the K calculated on the basis of the relevant formulas (4.19), (4.20) or (4.21), (4.22), which are given in Sect. 4.1.1.

It should be noted that—apart from the spiral curve—other transition curves may be used in horizontal alignment. These curves may be also used to forming the curvilinear geometric systems shown in Fig. 8.1. The above formulas (8.1)–(8.11) can also be used to determine and laying out the *Bloss* curve or other curves which are described using the curvature function. This of course requires a calculation of X_K and Y_K on the basis of the appropriate equations for these curves.

It can be added that similar design procedures are also possible in the case of the use of transition curves described using an explicit function. Selected proposals are described later in this chapter.

8.2 Parabolic Transition Curve as a Connecting Element Between Straight Line and Circular Arc

When using the parabolic transition curves, which are described in Sect. 5.1, notations u_K, x_K and y_K will be used instead of $\hat{u}$, X_K and Y_K (according to Figs. 5.1 and 7.1). The particular geometric elements are described using following formulas which follow from Fig. 8.1:

$$X_S = x_K - R_K \sin u_K = R_K \left[(n-1) \frac{\tan u_K}{\left(1 + \tan^2 u_K\right)^{3/2}} - \sin u_K \right] \tag{8.12}$$

$$H = y_K - R_K(1 - \cos u_K) = R_K \left[\frac{n-1}{n} \frac{\tan^2 u_K}{\left(1 + \tan^2 u_K\right)^{3/2}} + \cos u_K - 1 \right] \tag{8.13}$$

$$Y_S = R_K + H = R_K \left[\frac{n-1}{n} \frac{\tan^2 u_K}{\left(1 + \tan^2 u_K\right)^{3/2}} + \cos u_K \right] \tag{8.14}$$

$$T = x_K + y_K \tan u_K = R_K (n-1) \frac{\tan u_K}{\left(1 + \tan^2 u_K\right)^{3/2}} \left[1 + \frac{\tan^2 u_K}{n} \right] \tag{8.15}$$

$$T_d = x_K - y_K \frac{1}{\tan u_K} = R_K \frac{(n-1)^2}{n} \frac{\tan u_K}{\left(1 + \tan^2 u_K\right)^{3/2}} \tag{8.16}$$

$$T_k = y_K \frac{1}{\sin u_K} = R_K \frac{n-1}{n} \frac{\tan^2 u_K}{\left(1 + \tan^2 u_K\right)^{3/2}} \frac{1}{\sin u_K} \tag{8.17}$$

$$N = y_K \frac{1}{\cos u_K} = R_K \frac{n-1}{n} \frac{\tan^2 u_K}{\left(1 + \tan^2 u_K\right)^{3/2}} \frac{1}{\cos u_K} \tag{8.18}$$

$$U = y_K \frac{1}{\tan u_K} = R_K \frac{n-1}{n} \frac{\tan u_K}{\left(1 + \tan^2 u_K\right)^{3/2}} \tag{8.19}$$

$$V = Y_K \tan u_K = R_K \frac{n-1}{n} \frac{\tan^3 u_K}{\left(1 + \tan^2 u_K\right)^{3/2}} \tag{8.20}$$

$$C = \sqrt{X_K^2 + Y_K^2} = R_K(n-1) \frac{\tan u_K}{\left(1 + \tan^2 u_K\right)^{3/2}} \sqrt{1 + \frac{\tan^2 u_K}{n^2}} \tag{8.21}$$

$$\omega = \arctan \frac{y_K}{x_K} \tag{8.22}$$

whereby the coordinates x_K and y_K of the K are described by Eqs. (5.18) and (5.19).

8.3 Sinusoid as Transition Curve Between a Straight Line and Circular Arc

When using the sinusoid described by Eq. (5.25), general principles for determining the curvilinear transition between straight line and circular arc would be similar to those used in the case of the spiral curve. The main difference lies in the appropriate approach for calculating the necessary geometric elements, if they would be referenced to the main tangent, i.e. the tangent at the start point P (as in the case spiral curve) (Fig. 8.2). Equations necessary to determine the sinusoid as a transition curie between the straight line and the circular arc were presented in the article (Kobryń 2006).

Following formulas can be written on the basis of Fig. 8.2:

$$T = \frac{x_K}{\cos u_P} = R_K \frac{\Pi \tan u_P}{2 \cos u_P} \tag{8.23}$$

$$T_d = \frac{y_K}{\sin u_P} = R_K \frac{\tan u_P}{\cos u_P} \tag{8.24}$$

$$N = T \sin u_P - y_K = R_K \left(\frac{\Pi}{2} - 1 \right) \tan^2 u_P \tag{8.25}$$

$$T_k = \frac{N}{\tan u_P} = R_K \left(\frac{\Pi}{2} - 1 \right) \tan u_P \tag{8.26}$$

$$X_K = T - N \sin u_P = R_K \left[\frac{\Pi \tan u_P}{2 \cos u_P} - \left(\frac{\Pi}{2} - 1 \right) \tan^2 u_P \sin u_P \right] \tag{8.27}$$

$$Y_K = N \cos u_P = R_K \left(\frac{\Pi}{2} - 1 \right) \tan^2 u_P \cos u_P \tag{8.28}$$

$$H = (N + R_K) \cos u_P - R_K = R_K \left[\left(\frac{\Pi}{2} - 1 \right) \tan^2 u_P \cos u_P + \cos u_P - 1 \right] \tag{8.29}$$

$$X_S = X_K - R_K \sin u_P = R_K \left[\frac{\Pi \tan u_P}{2 \cos u_P} - \left(\frac{\Pi}{2} - 1 \right) \tan^2 u_P \sin u_P - \sin u_P \right] \tag{8.30}$$

$$Y_S = H + R_K = R_K \left[\left(\frac{\Pi}{2} - 1 \right) \tan^2 u_P \cos u_P + \cos u_P \right] \tag{8.31}$$

$$U = T_k \cos u_P = R_K \left(\frac{\Pi}{2} - 1 \right) \sin u_P \tag{8.32}$$

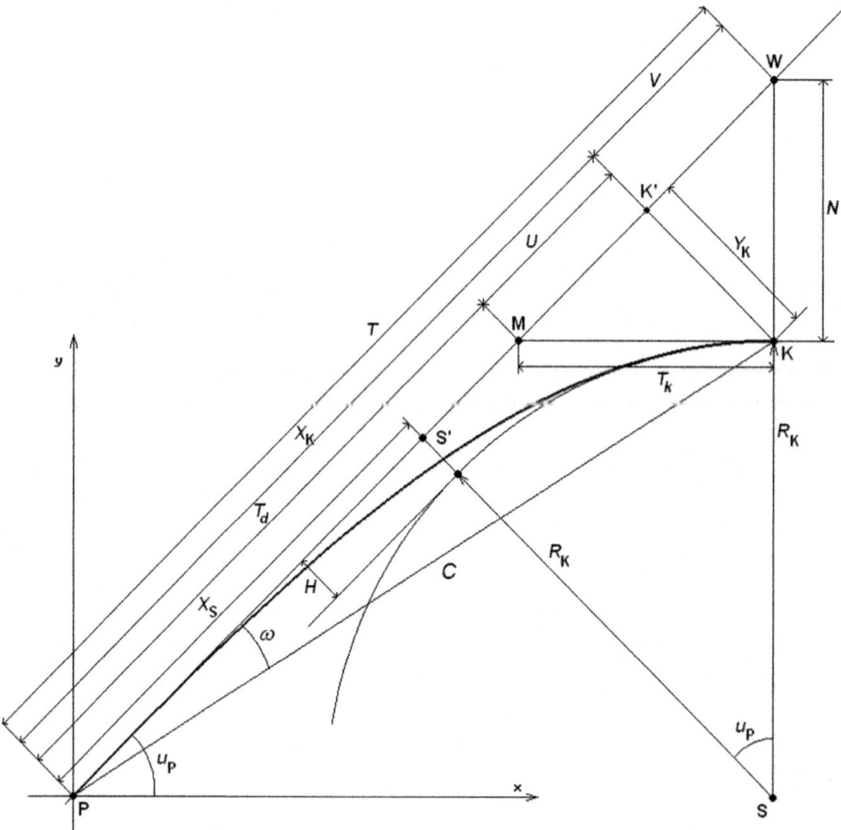

Fig. 8.2 Sinusoid as transition curve between a straight line and circular arc

$$V = N \sin u_P = R_K \left(\frac{\Pi}{2} - 1 \right) \tan^2 u_P \sin u_P \tag{8.33}$$

$$C = \sqrt{x_K^2 + y_K^2} = \frac{R_K \tan u_P}{2} \sqrt{\Pi^2 + 4 \tan^2 u_P} \tag{8.34}$$

$$\omega = \arctan \frac{Y_K}{X_K} \tag{8.35}$$

Whereas, to determine of the intermediate points of the sinusoid, it can be used formulas which define positions of these points with respect to the tangent at the point P.

According to Fig. 8.2, coordinates x' and y' can be calculated using the following transformation formulas:

$$x' = x \cos u_P + y \sin u_P \tag{8.36}$$

and

$$y' = x \sin u_P - y \cos u_P \tag{8.37}$$

where $t \in \langle 0; 1 \rangle$ and ordinate y is calculated on the basis of Eq. (5.27):

$$y = \frac{2x_K \tan u_P}{\Pi} \sin\left(\frac{\Pi}{2} t\right)$$

8.4   Polynomial Transition Curve as Connecting Element Between a Straight Line and Circular Arc

On similar principles as for the sinusoid, some solutions of the polynomial transition curves can be used to design of curvilinear transition between straight line and circular arc (Kobryń 2017). These solutions are presented in Chap. 7. These are the curves TC_HT-K_GS-1 i TC_HT-K_GS-3, which equations can be written in a more comfortable form (Kobryń 2002, 2011):

- for curves TC_HT-K_GS-1

$$y = \frac{x_K^2}{R_K}\left[Et + \frac{1 - 3E}{3} t^3 - \frac{1 - 2E}{4} t^4\right] \tag{8.38}$$

whereby $E \in \langle 1/3; 2/3 \rangle$

- for curves TC_HT-K_GS-3

$$y = \frac{x_K^2}{R_K}\left[Et + \frac{2 - 5E}{2} t^4 - \frac{7 - 15E}{5} t^5 + \frac{1 - 2E}{2} t^6\right] \tag{8.39}$$

whereby $E \in <4/10; 6/10 >$

In the case of the curves (8.38) and (8.39), a following relationship plays an important role in the design (Kobryń 2002, 2011):

$$x_K = \frac{R_K \tan u_P}{E} \tag{8.40}$$

An ordinate of the end point K follows from Eqs. (8.38) and (8.39) for $x = x_K$ (i.e. for $t = 1$) and is described using following equations:

- for curves (8.38)

$$y_K = R_K \tan^2 u_P \frac{6E + 1}{12E^2} \tag{8.41}$$

- for curves (8.39)

$$y_K = R_K \tan^2 u_P \frac{5E + 1}{10E^2} \tag{8.42}$$

Formulas for calculating other geometric elements result from Fig. 8.2, where appropriate designations are the same both for the sinusoid as well as for polynomial transition curves. According to (Kobryń 2017), a summary of these formulas are shown in Table 8.1 for the curves (8.38) and in Table 8.2 for the curves (8.39).

To determine the intermediate points, Eq. (8.38) or (8.39) can be used and then the transformation formulas (8.36) and (8.37) can be applied.

Table 8.1 Formulas describing elements necessary for laying out a horizontal arc with curves (8.38)

Geometrical element	Formula
Main tangent	$T = R_K \dfrac{\tan u_P}{\cos u_P} \dfrac{1}{E}$
Long tangent	$T_d = R_K \dfrac{\tan u_P}{\cos u_P} \dfrac{6E + 1}{12E^2}$
Normal	$N = R_K \tan^2 u_P \dfrac{6E - 1}{12E^2}$
Short tangent	$T_k = R_K \tan u_P \dfrac{6E - 1}{12E^2}$
Abscissa of the K relative to the main tangent	$X_K = R_K \dfrac{\tan u_P}{\cos u_P} \left(\dfrac{1}{E} - \sin^2 u_P \dfrac{6E - 1}{12E^2} \right)$
Ordinate of the K relative to the main tangent	$Y_K = R_K \tan^2 u_P \cos u_P \dfrac{6E - 1}{12E^2}$
Offset of the offsetted point of curvature	$H = R_K \left[\cos u_P \left(\tan^2 u_P \dfrac{6C - 1}{12C^2} + 1 \right) - 1 \right]$
Abscissa of the centre S	$X_S = R_K \sin u_P \left[\dfrac{1}{\cos^2 u_P} \left(\dfrac{1}{E} - \sin^2 u_P \dfrac{6E - 1}{12E^2} \right) - 1 \right]$
Ordinate of the centre S	$Y_S = R_K \cos u_P \left(\tan^2 u_P \dfrac{6E - 1}{12E^2} + 1 \right)$
Subtangent	$U = R_K \sin u_P \dfrac{6E - 1}{12E^2}$
Subnormal	$Y = R_K \sin u_P \tan^2 u_P \dfrac{6E - 1}{12E^2}$
Chord	$C = R_K \tan u_P \sqrt{ \dfrac{1}{E^2} + \tan^2 u_P \left(\dfrac{6E + 1}{12E^2} \right)^2 }$

Table 8.2 Formulas describing elements necessary for laying out a horizontal arc with curves (8.39)

Geometrical element	Formula
Main tangent	$T = R_K \frac{\tan u_P}{\cos u_P} \frac{1}{E}$
Long tangent	$T_d = R_K \frac{\tan u_P}{\cos u_P} \frac{5E+1}{10E^2}$
Normal	$N = R_K \tan^2 u_P \frac{5E-1}{10E^2}$
Short tangent	$T_k = R_K \tan u_P \frac{5E-1}{10E^2}$
Abscissa of the K relative to the main tangent	$X = R_K \frac{\tan u_P}{\cos u_P} \left(\frac{1}{E} - \sin^2 u_P \frac{5E-1}{10E^2} \right)$
Ordinate of the K relative to the main tangent	$Y = R_K \tan^2 u_P \cos u_P \frac{5E-1}{10E^2}$
Offset of the offsetted point of curvature	$H = R_K \left[\cos u_P \left(\tan^2 u_P \frac{5E-1}{10E^2} + 1 \right) - 1 \right]$
Abscissa of the centre S	$X_S = R_K \sin u_P \left[\frac{1}{\cos^2 u_P} \left(\frac{1}{E} - \sin^2 u_P \frac{5E-1}{10E^2} \right) - 1 \right]$
Ordinate of the centre S	$Y_S = R_K \cos u_P \left(\tan^2 u_P \frac{5E-1}{10E^2} + 1 \right)$
Subtangent	$U = R_K \sin u_P \frac{5E-1}{10E^2}$
Subnormal	$Y = R_K \sin u_P \tan^2 u_P \frac{5E-1}{10E^2}$
Chord	$C = R_K \tan u_P \sqrt{ \frac{1}{E^2} + \tan^2 u_P \left(\frac{5E+1}{10E^2} \right)^2 }$

8.5   General Transition Curves as Connecting Element Between Two Straight Lines

Use of the general transitions curves for the design of horizontal curves may allow the description of the entire curvilinear transition between two straight lines using only one equation (Fig. 8.3).

For this purpose, either the sinusoid (5.43), as well as various solutions of the general transition curves can be used, which are presented in Chap. 7. In this way, it can be designed a horizontal arcs, which are equivalent to conventional geometrical systems in the form of:

- 1st transition curve—circular arc—2nd transition curve
- 1st transition curve—2nd transition curve

In the case of the sinusoid given by Eq. (5.43), design of symmetrical arcs is possible, whereas various types of arcs can be designed in the case of general transition curves (symmetric or unsymmetric).

8.5.1   Designing Horizontal Curves Using Sinusoid as a General Transition Curve

Laying out of the sinusoid given by Eq. (5.43) as a general transition curve requires knowledge about length of tangents $\overline{PW}$ and $\overline{KW}$ (Fig. 8.3). On the basis of Fig. 8. 3, it follows:

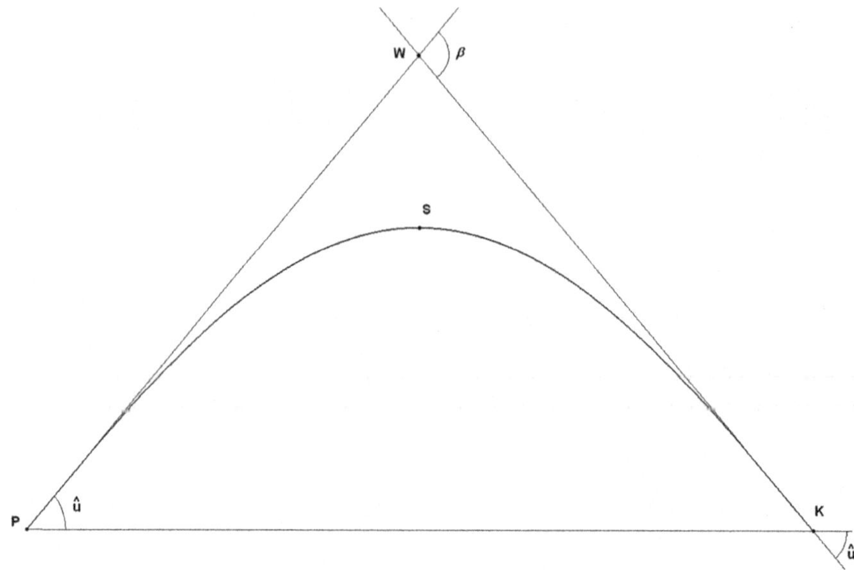

Fig. 8.3 General transition curve as curvilinear transition between two straight lines

$$\overline{PW} = \overline{KW} = x_K \frac{\sin \hat{u}}{\sin \beta} \tag{8.43}$$

whereby β is deflection angle that follows from equation:

$$\beta = 180° - 2\hat{u} \tag{8.44}$$

From Eq. (5.43) follows that an ordinate of the middle point S (for $x_S = 0.5 \cdot x_K$) is $y_S = \frac{x_K \tan \hat{u}}{\Pi}$.

It can be noted that the laying out of the entire arc of the curve (5.43) with respect to tangents $\overline{PW}$ and $\overline{KW}$ requires a knowledge of the coordinates of only half arc $\overset{\frown}{PS}$. The other half $\overset{\frown}{SK}$ is symmetric reflection with respect to the point S. In the case of setting out of the sinusoid (8.43) relative to the tangents $\overline{PW}$ and $\overline{KW}$ can be used transformation formulas, which are analogous to Eqs. (8.36) and (8.37). For this purpose, the angle $\hat{u}$ and values x and y which relate to the a half curve, i.e. to the section $\overset{\frown}{PS}$, should be used in these equations.

8.5.2   Designing Horizontal Curves
Using Polynomial General Transition Curves

The general transition curves presented in the paper (Kobryń 2014a) are well suited for the design of horizontal curves according to Fig. 8.3. Among other, these are curves with the identification code GSSTC_GS-2 and GSSTC_GS-4, which were presented in Sect. 7.4. Their basis equations, which are given in Tables 7.8 and 7.16, are in the form:

$$y = x_K \tan \alpha \cdot M_0 + x_K \tan u_P \cdot M_1 + x_K \tan u_K \cdot M_2 \tag{8.45}$$

and

$$y = x_K \tan \alpha \cdot G_0 + x_K \tan u_P \cdot G_1 + x_K \tan u_K \cdot G_2 \tag{8.46}$$

Assuming $\tan \alpha = 0$, these equations can be written as:

$$y = x_K (\tan u_P \cdot M_1 + \tan u_K \cdot M_2) \tag{8.47}$$

where:

$$M_1 = t - 6t^3 + 8t^4 - 3t^5$$
$$M_2 = -4t^3 + 7t^4 - 3t^5$$

and

$$y = x_K (\tan u_P \cdot G_1 + \tan u_K \cdot G_2) \tag{8.48}$$

where:

$$G_1 = t - 20t^4 + 45t^5 - 36t^6 + 10t^7$$
$$G_2 = -15t^4 + 39t^5 - 34t^6 + 10t^7$$

In the case, when $\tan u_P = -\tan u_K = \tan \hat{u}$, it results:

- from Eq. (8.47)

$$y = x_K \tan \hat{u} (t - 2t^3 + t^4) \tag{8.49}$$

- from Eq. (8.48)

$$y = x_K \tan \hat{u}\left(t - 5t^4 + 6t^5 - 2t^6\right) \tag{8.50}$$

From Eqs. (8.49) and (8.50) results that an abscissa of the arc centre (for $x_S = 0,5 \cdot x_K$) is:

- $y_S = 0.8125 x_K \tan \hat{u}$ for the curves (8.49)
- $y_S = 0.84375 x_K \tan \hat{u}$ for the curves (8.50)

General principles for designing of horizontal curve using the polynomial transition curves will be similar as in the case of the sinusoid. The laying out of the whole arc of the curve relative to the tangents $\overline{PW}$ and $\overline{KW}$ requires a knowledge only about the coordinates half arc $\left(\overset{\frown}{PS}\right)$, since the second half $\left(\overset{\frown}{SK}\right)$ is a symmetrical reflection with respect to the point S. To determine the intermediate points relative to the tangents $\overline{PW}$ and $\overline{KW}$, the transformation Eqs. (7.36) and (7.37) should be used. For this purpose, an angle $\hat{u}$ and values of x and y which relate to half arc, i.e. the section $\overset{\frown}{PS}$, should be used.

8.6 Universal Transition Curves in Horizontal Alignment

The paper (Kobryń 2016b) describes a design principles for curvilinear transition between any two points by the use of the universal transition curves, which are described in Sect. 7.4. It is about the curves with the identification codes UOTC_GS-1 and UOTC_GS-2. It can be recalled that the equations of these curves have a form:

- curves UOTC_GS-1

$$\begin{aligned}
y = x_K \big[N_1 \tan u_P + N_2 \tan u_K \\
+ n_P N_3 \frac{x_K}{R_P}\left(1 + \tan^2 u_P\right)^{3/2} + n_K N_4 \frac{x_K}{R_K}\left(1 + \tan^2 u_K\right)^{3/2} \big]
\end{aligned} \tag{8.51}$$

where:

$$N_1 = t - t^3 + \frac{1}{2}t^4$$

$$N_2 = t^3 - \frac{1}{2}t^4$$

$$N_3 = \frac{1}{2}t^2 - \frac{2}{3}t^3 + \frac{1}{4}t^4$$

$$N_4 = -\frac{1}{3}t^3 + \frac{1}{4}t^4$$

- curves UOTC_GS-2

$$y = x_K \left[M_0 \tan \alpha + M_1 \tan u_P + M_2 \tan u_K \right.$$
$$\left. + m_P M_3 \frac{x_K}{R_P} \left(1 + \tan^2 u_P\right)^{3/2} + m_K M_4 \frac{x_K}{R_K} \left(1 + \tan^2 u_K\right)^{3/2} \right] \tag{8.52}$$

where:

$$M_0 = 10t^3 - 15t^4 + 6t^5$$
$$M_1 = t - 6t^3 + 8t^4 - 3t^5$$
$$M_2 = -4t^3 + 7t^4 - 3t^5$$
$$M_3 = \frac{1}{2}t^2 - \frac{3}{2}t^3 + \frac{3}{2}t^4 - \frac{1}{2}t^5$$
$$M_4 = \frac{1}{2}t^3 - t^4 + \frac{1}{2}t^5$$

8.6.1   Designing of Curvilinear Transitions Using First Solution of Universal Transition Curves

In the case of the curves (8.51), it should be noted that the ordinate y_K of the end point (for $x = x_K$, i.e. $t = 1$) is described by a following equation:

$$y_K = x_K \left[\frac{1}{2} \tan u_P + \frac{1}{2} \tan u_K \right.$$
$$\left. + n_P \frac{1}{12} \frac{x_K}{R_P} \left(1 + \tan^2 u_P\right)^{3/2} - n_K \frac{1}{12} \frac{x_K}{R_K} \left(1 + \tan^2 u_K\right)^{3/2} \right] \tag{8.53}$$

Eq. (8.53) will be useful in the design procedure, because it defines a relationship between the values x_K, $\tan u_P$, $\tan u_K$, R_P, R_K and y_K. Furthermore, the derivatives y', y'' and y'' of the function (8.51) are used:

- derivative y'

$$y' = N_1' \tan u_P + N_2' \tan u_K +$$
$$+ n_P N_3' \frac{x_K}{R_P} \left(1 + \tan^2 u_P\right)^{3/2} + n_K N_4' \frac{x_K}{R_K} \left(1 + \tan^2 u_K\right)^{3/2} \tag{8.54}$$

where:

$$N_1' = 1 - 3t^2 + 2t^3$$
$$N_2' = 3t^2 - 2t^3$$
$$N_3' = t - 2t^2 + t^3$$
$$N_4' = -t^2 + t^3$$

- derivative y''

$$y'' = N_1'' \frac{1}{x_K} \tan u_P + N_2'' \frac{1}{x_K} \tan u_K + $$
$$+ n_P N_3'' \frac{1}{R_P} \left(1 + \tan^2 u_P\right)^{3/2} + n_K N_4'' \frac{1}{R_K} \left(1 + \tan^2 u_K\right)^{3/2} \qquad (8.55)$$

where:

$$N_1'' = -6t + 6t^2$$
$$N_2'' = 6t - 6t^2$$
$$N_3'' = 1 - 4t + 3t^2$$
$$N_4'' = -2t + 3t^2$$

- derivative y''

$$y''' = N_1''' \frac{1}{x_K^2} \tan u_P + N_2''' \frac{1}{x_K^2} \tan u_K + $$
$$+ n_P N_3''' \frac{1}{x_K R_P} \left(1 + \tan^2 u_P\right)^{3/2} + n_K N_4''' \frac{1}{x_K R_K} \left(1 + \tan^2 u_K\right)^{3/2} \qquad (8.56)$$

where:

$$N_1''' = -6 + 12t$$
$$N_2''' = 6 - 12t$$
$$N_3''' = -4 + 6t$$
$$N_4''' = -2 + 6t$$

Two variants of the curve geometry are possible:

- existence of one inflection point (Fig. 7.7),
- lack of inflection points (Fig. 7.8).

A basic condition for the existence of a single inflection point within the curve (8.51) are opposed bulges at the points P and K, which means:

$$n_P \cdot n_K = -1 \qquad (8.57)$$

Assuming the given positions of the points P and K (i.e. $P(X_P, Y_P)$ and $K(X_K, Y_K)$), we obtain from Fig. 8.4:

$$y_K = Y_K - Y_P \qquad (8.58)$$

$$x_K = X_K - X_P \qquad (8.59)$$

In addition to the coordinates x_K and y_K, we assume also that tangent orientations at the points P and K are known ($\tan u_P = \tan U_P$ and $\tan u_K = \tan U_K$). Then, after adoption one of the values R_P or R_K, the missing value (appropriately R_K or R_P) can be calculated on the basis of Eq. (8.53).

The curvature within the arc formed by the curve (8.51) should not exceed a designed maximal value $1/R_{\min}$. For this purpose, a location of points must be identified wherein exists a extremum of the curvature. It is possible on the basis of a necessary condition (7.4), wherein the derivatives y', y'' and y''' will be expressed by the use of Eqs. (8.54), (8.55) and (8.56). To solve Eq. (7.4) with respect to variable t we need to know the values of x_K, $\tan u_P$, $\tan u_K$, R_P and R_K.

In the interval $t \in\ <0; 1>$, Eq. (7.4) could have one or two solutions which describe position of the point (points), where exists the maximal curvature:

t_E in the case of absence the inflection points within the curve (8.51),

$t_E^{(P)}$ and/or $t_E^{(K)}$ in the case of existence the inflection point within the curve (8.52), (notation (P) or (K) means a position of the maximal curvature appropriately at initial or end part of the curve).

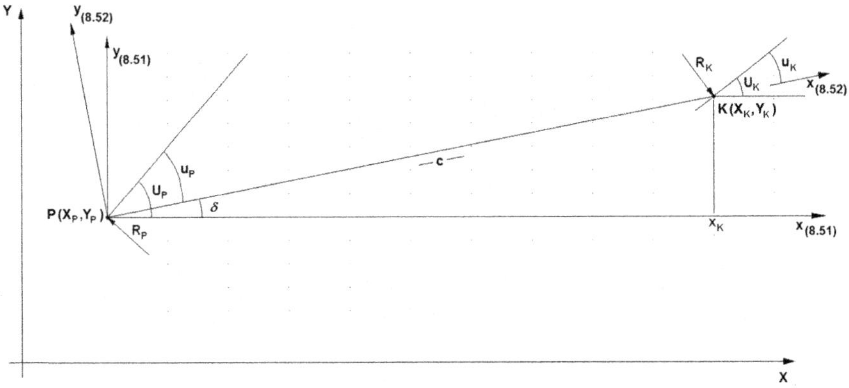

Fig. 8.4 Designing of horizontal curves using the universal transition curves (*with permission from ASCE*)

Assuming $t = t_E$ (or $t = t_E^{(P)}$ and/or $t = t_E^{(K)}$) and appropriately expressing the derivatives y' and y'' by the use of Eqs. (8.54) and (8.55), the appropriate values of the curvature can be calculated on the basis Eq. (7.2):

k_E — in the case of absence the inflection points within the curve (8.51),

$k_E^{(P)}$ and/or $k_E^{(K)}$ — in the case of existence the inflection point within the curve (8.51).

Values of the curvature calculated in this way shall not exceed the designed maximal value $1/R_{\min}$.

8.6.2 Designing of Curvilinear Transitions Using Second Solution of Universal Transition Curves

In the case of the curves (8.52), it results from their equation for $x = x_K$ (i.e. $t = 1$):

$$y_K = x_K \tan \alpha \tag{8.60}$$

Similarly, as for the curves (8.51), the further calculation procedure uses the following derivatives of the function (8.52). The derivatives y', y'' and y''' are:

- derivative y'

$$y' = M_0' \tan \alpha + M_1' \tan u_P + M_2' \tan u_K +$$
$$+ m_P M_3' \frac{x_K}{R_P} \left(1 + \tan^2 u_P\right)^{3/2} + m_K M_4' \frac{x_K}{R_K} \left(1 + \tan^2 u_K\right)^{3/2} \tag{8.61}$$

where:

$$M_0' = 30t^2 - 60t^3 + 30t^4$$
$$M_1' = 1 - 18t^2 + 32t^3 - 15t^4$$
$$M_2' = -12t^2 + 28t^3 - 15t^4$$
$$M_3' = t - \frac{9}{2}t^2 + 6t^3 - \frac{5}{2}t^4$$
$$M_4' = \frac{3}{2}t^2 - 4t^3 + \frac{5}{2}t^4$$

- derivative y''

$$y'' = M_0'' \frac{1}{x_K} \tan \alpha + M_1'' \frac{1}{x_K} \tan u_P + M_2'' \frac{1}{x_K} \tan u_K +$$
$$+ m_P M_3'' \frac{1}{R_P} \left(1 + \tan^2 u_P\right)^{3/2} + m_K M_4'' \frac{1}{R_K} \left(1 + \tan^2 u_K\right)^{3/2} \tag{8.62}$$

where:

$$M_0'' = 60t - 180t^2 + 120t^3$$
$$M_1'' = -36t + 96t^2 - 60t^3$$
$$M_2'' = -24t + 84t^2 - 60t^3$$
$$M_3'' = 1 - 9t + 18t^2 - 10t^3$$
$$M_4'' = 3t - 12t^2 + 10t^3$$

- derivative y'''

$$y''' = M_0''' \frac{1}{x_K} \tan \alpha + M_1''' \frac{1}{x_K^2} \tan u_P + M_2''' \frac{1}{x_K^2} \tan u_K +$$
$$+ m_P M_3''' \frac{1}{x_K R_P} \left(1 + \tan^2 u_P\right)^{3/2} + m_K M_4''' \frac{1}{x_K R_K} \left(1 + \tan^2 u_K\right)^{3/2} \tag{8.63}$$

where:

$$M_0'' = 60 - 360t + 360t^2$$
$$M_1'' = -36 + 192t - 180t^2$$
$$M_2'' = -24 + 168t - 180t^2$$
$$M_3'' = -9 + 36t - 30t^2$$
$$M_4'' = 3 - 24t + 30t^2$$

Due to the more extensive equation of the curve (8.52), an additional condition $(\tan \alpha = 0)$ can be assumed. This allows you to opt out of the first member of Eq. (8.52). According to Fig. 8.4, this means that the origin of the local coordinate system is at the point P, and the abscissa axis passes through the point K. Therefore it is $x_K = c$, whereby:

$$c = \sqrt{(X_K - X_P)^2 + (Y_K - Y_P)^2} \tag{8.64}$$

The inclination angle of the chord connecting the points P and K in the super-ordinated system is:

$$\delta = \arctan \frac{Y_K - Y_P}{X_K - X_P} \tag{8.65}$$

It follows that the angles of the tangent inclinations at the points P and K in the local coordinate system are:

$$u_P = U_P - \delta \tag{8.66}$$

$$u_K = U_K - \delta \tag{8.67}$$

Just like in Sect. 8.5.1, two geometric variants also in this case are possible:

- existence the inflection point,
- absence the inflection points.

Condition for the existence of one inflection point within the curve (8.51) has a form:

$$m_P \cdot m_K = -1 \tag{8.68}$$

Further procedure is similar to that described in Sect. 8.5.1. The curvature within the arc formed by the curve (8.52) should not exceed the designed maximal value $1/R_{\min}$. To determine the location of points in which there is extreme curvature, a precondition (7.4) may be used, wherein the derivatives y', y'' and y''' will be expressed by the use of Eqs. (8.61), (8.62) and (8.63). To solve Eq. (7.4) with respect to variable t we need to known the values of x_K, $\tan u_P$, $\tan u_K$, R_P and R_K. The value $x_K = c$ results from Eq. (8.64), $\tan u_P$ and $\tan u_K$ are obtained from (8.66) and (8.67).

Eq. (6.4) can have one or two solutions in the interval $t \in <0; 1>$. These solutions describe a position of the point (points) in which is maximal value of the curvature:

t_E in the case of absence the inflection points within the curve (8.53),

$t_E^{(P)}$ and/or $t_E^{(K)}$ in the case of existence the inflection point within the curve (8.52) ((P) or (K) means a position of the maximal curvature appropriately in begin or final part of the curve).

Assuming $t = t_E$ (or $t = t_E^{(P)}$ and/or $t = t_E^{(K)}$) and appropriately expressing the derivatives y' and y'' by the use of Eqs. (8.61) and (8.62), the appropriate values of the curvature can be calculated on the basis Eq. (7.2):

k_E in the case of an absence the inflection points within the curve (8.52),

$k_E^{(P)}$ and/or $k_E^{(K)}$ in the case of an existence the inflection point within the curve (8.52).

Values of the curvature calculated in this way shall not exceed the designed maximal value $1/R_{\min}$.

References

Brockenbrough RL (ed) (2009) Highway engineering handbook, 3rd edn. McGraw-Hill, Professional Book Group, New York

Easa SM (2003) Geometric design. In: Chen WF, Liew JYR (eds) The civil engineering handbook. CRC Press, Taylor & Francis Group, Boca Raton

Kobryń A (2002) Wielomianowe krzywe przejściowe w projektowaniu niwelety tras drogowych. Wydawnictwa Politechniki Białostockiej, Rozprawy Naukowe nr 100, Białystok (in Polish)

Kobryń A (2006) Sinusoida jako krzywa przejściowa. Drogownictwo 61(7):242–245 (in Polish)

Kobryń A (2011) Polynomial solutions of transition curves. J. Surv. Eng. 137(3):71–80

Kobryń A (2014) New solutions for general transition curves. J. Surv. Eng. 140(1):12–21

Kobryń A (2016) Universal solutions of transition curves. J. Surv. Eng. 142(4):1–16

Kobryń A (2017) Use of polynomial transition curves in the design of horizontal arcs. Roads and Bridges 16(1):5–14

Lamm R, Psarianos B, Mailänder T (1999) Highway design and traffic safety engineering handbook. McGraw-Hill, Professional Book Group, New York

Lipiński M (1993) Geometria i tyczenie tras drogowych. In: Geodezja inżynieryjna, tom 3. Polskie Przedsiębiorstwo Wydawnictw Kartograficznych, Warszawa – Wrocław (in Polish)

Lorenz H (1971) Trassierung und Gestaltung vion Strassen und Autobahnen. Wiesbaden – Berlin (in German)

Meyer CF, Gibson DW (1980) Route surveying and design. Harper & Row, New York

Rogers M (2008) Highway engineering, 2nd edn. Wiley-Blackwell, Chichester-Oxford

Wolhuter KM (2015) Geometric design of roads handbook. CRC Press, Taylor & Francis Group, Boca Raton

Chapter 9
Sample Applications of Transition Curves in Vertical Alignment

9.1 Optimization of Vertical Alignment Using Polynomial Transition Curves

Optimization of vertical alignment is mainly aimed at minimizing earthworks. This is achieved by adjusting the vertical alignment to the longitudinal terrain profile. For this purpose, the polynomial transition curves may be also used, which are described in Sect. 7.4. In a further part of Sect. 9.1 were described procedures for the use of two families of curves with horizontal tangent at the end point (TC_HT-K_GS-1, TC_HT-K_GS-3) and fourth families of general transition curves (GTC_GS-1, GSSTC_GS-2, GTC_GS-3 and GSSTC_GS-4).

9.1.1 Optimization of Vertical Alignment Using Polynomial Transition Curves with Horizontal Tangent at End Point

Optionally, the equations of the transition curves TC_HT-K_GS-1 and TC_HT-K_GS-3 can be written in a more convenient form:

- for the curves TC_HT-K_GS-1 (see Sect. 7.4)

$$y = \frac{x_K \tan u_P}{E}\left[Et + \frac{1-3E}{3}t^3 - \frac{1-2E}{4}t^4\right] \tag{9.1}$$

whereby $E \in <1/3; 2/3>$
- for the curves TC_HT-K_GS-3 (see Sect. 7.4)

© Springer International Publishing AG 2017

A. Kobryń, *Transition Curves for Highway Geometric Design*, Springer Tracts on Transportation and Traffic 14, DOI 10.1007/978-3-319-53727-6_9

$$y = \frac{x_K \tan u_P}{E} \left[Et + \frac{2-5E}{2} t^4 - \frac{7-15E}{5} t^5 + \frac{1-2E}{2} t^6 \right] \qquad (9.2)$$

whereby $E \in\ <4/10;\ 6/10>$

The possibility of choosing the parameter E within the range $E \in\ <1/3;\ 2/3>$ or $E \in\ <4/10;\ 6/10>$ gives a great possibilities to shape the curves geometry. This is illustrated on Fig. 9.1, which shows graphs of these curves for different values of the parameter E (assuming the same value $\tan u_P$ and R_K).

According to the work (Kobryń 2002), the use of these curves in optimization of vertical alignment consists in connecting the subsequent curves so as to obtain smoothness of vertical alignment over the full length (Fig. 9.2).

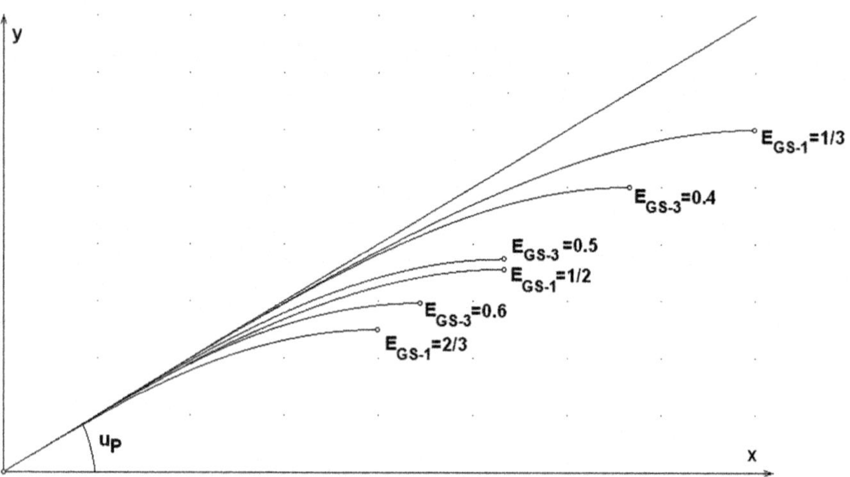

Fig. 9.1 Example graphs of curves (9.1) and (9.2) (*with permission from ASCE*)

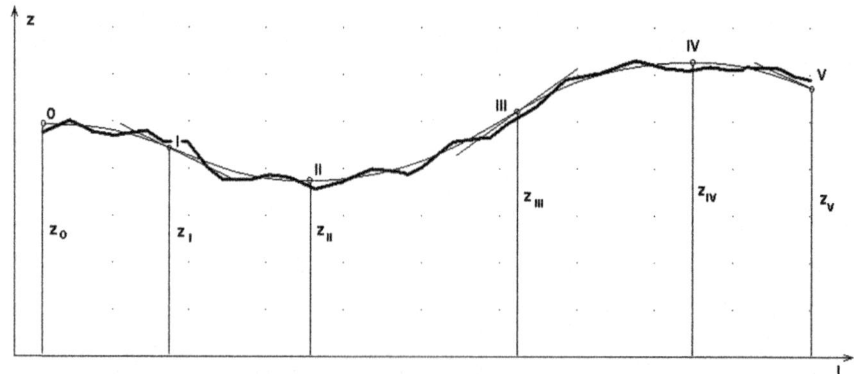

Fig. 9.2 General principle of vertical alignment designing using the transition curves

This requires a fulfillment of following conditions at the connection points between subsequent curves:

- compliance of heights,
- compliance of tangent inclinations.

Therefore, every second curve should be designed as the mirror image of the curve described by Eq. (9.1) or (9.2) in the normal position. This is illustrated in Figs. 9.3 and 9.4.

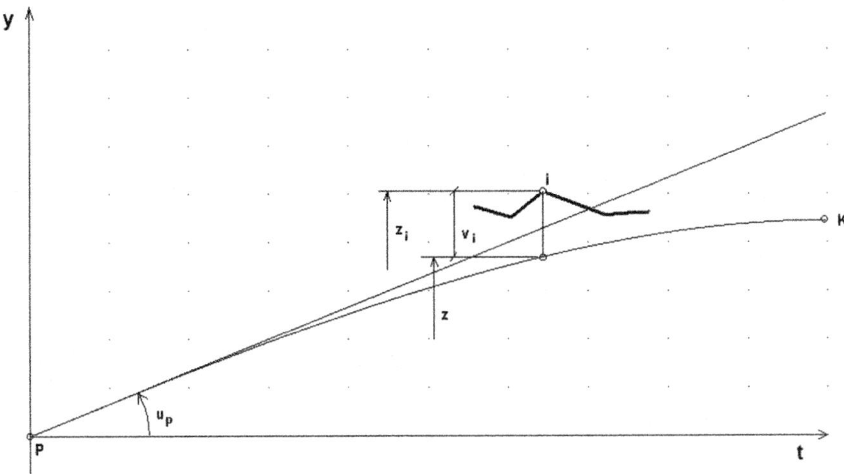

Fig. 9.3 Section of grade line formed by the curve in its original position

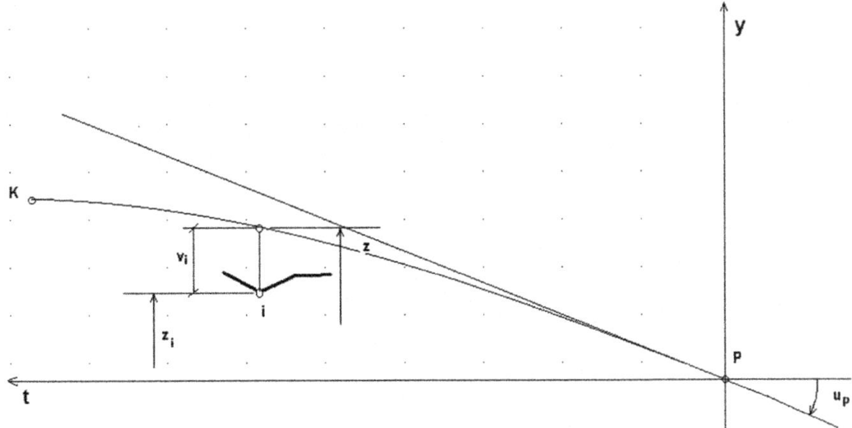

Fig. 9.4 Section of grade line formed by the curve in its mirror position

Designing of vertical alignment using the transition curves (9.1) and (9.2) comprises:

- establishment appropriate approximation equations in subsequent points of the longitudinal profile,
- processing the system of the approximation equations by the use least squares method.

This leads to the determination of the unknowns, which are the tangents inclinations at the appropriate points of the curves connection. In the case shown in Fig. 9.2 it would be the points I and III. Tangent at the connecting points II and IV would be horizontal.

On Figs. 9.3 and 9.4 were assumed following designations:

z_i terrain height at i-th point of a longitudinal profile,
z grade line height at i-th point of a longitudinal profile,
v_i approximation correction at i-th point of a longitudinal profile, that is, the size which added to the ordinate z_i allows to calculate the height of the grade line at this point.

Moreover, L_i, L_P, L_K are the mileage of the points i, P and K.

According to Fig. 9.3, a grade line height at any i-th point of the longitudinal profile can be expressed as

$$z = z_i + v_i \tag{9.3}$$

The height of the grade line can be expressed also as

$$z = z_P + y_i \tag{9.4}$$

where:

z_P terrain height at the start point P of the curve forming a considered fragment of the grade line,
y_i an ordinate of i-th point in the local coordinates system, the beginning of which coincides with the initial point of the curve (Fig. 9.2).

It follows the following approximation equation:

$$v_i = y_i + z_P - z_i \tag{9.5}$$

After expressing the ordinate y_i by the use of appropriate equation of the curve, the following approximation equations were obtained:

- for curves (9.1)

$$v_i = \tan u_P \frac{x_K}{E}\left[Et_i + \frac{1 - 3E}{3}t_i^3 - \frac{1 - 2E}{4}t_i^4\right] + z_P - z_i \tag{9.6}$$

- for curves (9.2)

$$v_i = \tan u_P \frac{x_K}{E}\left[Et_i + \frac{2-5E}{2}t_i^4 - \frac{7-15E}{5}t_i^5 + \frac{1-2E}{2}t_i^6 \right] + z_P - z_i \qquad (9.7)$$

Variable t_i in Eqs. (9.6) and (9.7) is expressed as $t_i = (L_i - L_P)/(L_K - L_P)$.

Whereas, in the case as shown in Fig. 9.4, height of the grade line at i-th point of the longitudinal profile can be expressed using Eq. (9.3) and as

$$z = z_K + y_K - y_i \qquad (9.8)$$

It follows the approximation equation in the form:

$$v_i = y_K - y_i + z_K - z_i \qquad (9.9)$$

After expressing the ordinate y_i by the use of appropriate equation of the curve, the following approximation equations were obtained:

- for curves (9.1)

$$v_i = \tan u_P \frac{x_K}{E}\left[E(1 - t_i) + \frac{1-3E}{3}\left(1 - t_i^3\right) - \frac{1-2E}{4}\left(1 - t_i^4\right) \right] + z_K - z_i$$

$$(9.10)$$

- for curves (9.2)

$$v_i = \tan u_P \frac{x_K}{E}\left[E(1 - t_i) + \frac{2-5E}{2}\left(1 - t_i^4\right) - \frac{7-15E}{5}\left(1 - t_i^5\right) + \frac{1-2E}{2}\left(1 - t_i^6\right) \right]$$
$$+ z_K - z_i$$

$$(9.11)$$

Variable t_i in Eqs. (9.10) and (9.11) is expressed as $t_i = (L_P - L_i)/(L_P - L_K)$. Different variants of assumptions regarding the definition of the subsequent curves are acceptable, for example:

- value E can be the same for all curves,
- value E for each curve can be different,
- whole grade line can be built only by the curves (9.1) or curves (9.2),
- a particular sections of the grade line can be built by the use of any curves (9.1) and (9.2)

Points of the longitudinal profile can be located unevenly. Therefore, approximation equation should receive adequate weights. It is reasonable that weights are proportional to the distance between adjacent points, i.e. approximation equation for the i-th point of the longitudinal profile would receive a weight $p_i \approx L_{i+1} - L_{i-1}$.

After creating approximation equations for all points of the longitudinal profile yields the corresponding system of equations. Individual values in the approximation equations can be grouped and included in appropriate matrices:

$\mathbf{v}_{(m,1)}$ vector of the approximation corrections,

$\mathbf{u}_{(n,1)}$ vector of the calculated values $\tan u_P$,

$\mathbf{W}_{(m,n)}$ matrix of coefficients at values $\tan u_P$ in each approximation equations (whereby $m > n$),

$\mathbf{l}_{(m,1)}$ vector of free terms,

$\mathbf{P}_{(m,m)}$ weight matrix (positively determined).

A system of approximation equations has the form:

$$\mathbf{v} = \mathbf{W}\mathbf{u} + \mathbf{l} \tag{9.13}$$

It should be determined the values $\tan u_P$ at the connection points of subsequent curves. To do this, the equations system (9.13) should be solved using the method of least squares. On the basis of a condition

$$\mathbf{v}^{\mathbf{T}}\mathbf{P}\mathbf{v} = \min. \tag{9.14}$$

after differentiation and taking into account (9.13) is obtained:

$$\mathbf{W}^{\mathbf{T}}\mathbf{P}\mathbf{W}\mathbf{u} + \mathbf{W}^{\mathbf{T}}\mathbf{P}\mathbf{l} = \mathbf{0} \tag{9.15}$$

As a result of solution of the system (9.15) a vector of values $\tan u_P$ is obtained:

$$\mathbf{u} = -\mathbf{N}^{-1}\mathbf{M}. \tag{9.16}$$

where:

$\mathbf{N}_{(n,n)} = \mathbf{W}^{\mathbf{T}}\mathbf{P}\mathbf{W}$ (whereby rank of a matrix $R(\mathbf{N}) = n$) and $\mathbf{M}_{(n,1)} = \mathbf{W}^{\mathbf{T}}\mathbf{P}\mathbf{l}$.

After calculation the value of $\tan u_P$ a following checks are required:

- fulfillment of the limit values by the determined inclinations $\tan u_P$,
- fulfillment of the limit values by the minimum radii of curvature R_K, which for the particular values x_K and $\tan u_P$ can be calculated as:

$$R_K = \frac{E \cdot x_K}{\tan u_P} \tag{9.17}$$

9.1.2 Optimization of Vertical Alignment Using General Transition Curves

Similarly to curves analyzed in Sect. 9.1, also the general transition curves are a useful tool that can be used in the design of vertical alignment. The same principles can be used for all the solutions of general transition curves, which are defined in Sect. 7.4 (GTC_GS-1, GSSTC_GS-2, GTC_GS-3 and GSSTC_GS-4). Usefulness of these curves follows from this that they are the family of curves whose geometry and the curvature varies depending on relations between the values of $\tan u_P$, $\tan u_K$ and $\tan \alpha$ (if this value is present in the curve equation). This is illustrated in Figs. 9.5 and 9.6, which illustrate an exemplary graphs of curvature of the curves GSSTC_GS-2 and GSSTC_GS-4 (value t_E describes a location of curvature extremum).

Optimization of the vertical alignment using the general transition curves is presented in the works (Kobryń 2002, 2017). Similarly to the case described in Sect. 9.1.1, the design procedure consists in connecting subsequent curves so as to obtain smoothness vertical alignment over the full length. In the connection points between subsequent curves should therefore ensure:

- compliance of heights,
- compliance of tangent inclinations.

In accordance with Fig. 9.7 we can create equations identical to the Eqs. (9.3) and (9.4), from that follows the approximation Eq. (9.5). Expressing the ordinate y_i using equations of general transition curves, can be written

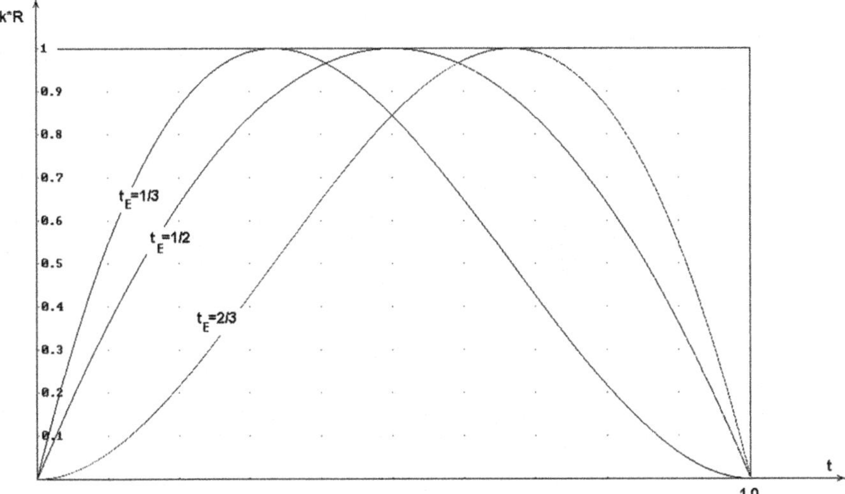

Fig. 9.5 Example graphs of curvature for the curves GSSTC_GS-2 (*with permission from ASCE*)

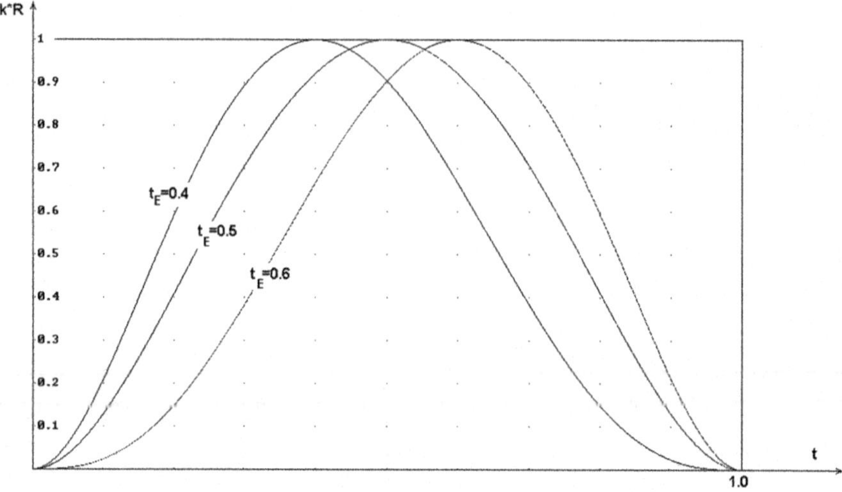

Fig. 9.6 Example graphs of curvature for the curves GSSTC_GS-4 (*with permission from ASCE*)

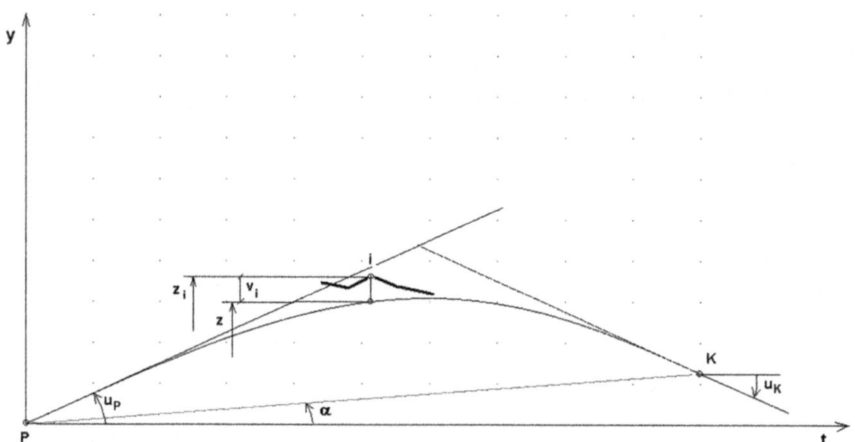

Fig. 9.7 Section of the grade line formed by the general transition curve

approximation equation for individual curves. According to article (Kobryń 2017), they have a form:

- for curves GTC_GS-1

$$v_i = x_K N_1^{(i)} \tan u_P + x_K N_2^{(i)} \tan u_K + z_P - z_i \qquad (9.18)$$

wherein $N_1^{(i)}$ and $N_2^{(i)}$ are expressed by appropriate dependencies for $t = t_i$;

- for curves GSSTC_GS-2

$$v_i = x_K M_0^{(i)} \tan\alpha + x_K M_1^{(i)} \tan u_P + x_K M_2^{(i)} \tan u_K + z_P - z_i \qquad (9.19)$$

wherein $M_0^{(i)}$, $M_1^{(i)}$ and $M_2^{(i)}$ are expressed by appropriate dependencies for $t = t_i$;

- for curves GTC_GS-3

$$v_i = x_K F_1^{(i)} \tan u_P + x_K F_2^{(i)} \tan u_K + z_P - z_i \qquad (9.20)$$

wherein $F_1^{(i)}$ and $F_2^{(i)}$ are expressed by appropriate dependencies for $t = t_i$;
- for curves GSSTC_GS-4

$$v_i = x_K G_0^{(i)} \tan\alpha + x_K G_1^{(i)} \tan u_P + x_K G_2^{(i)} \tan u_K + z_P - z_i \qquad (9.21)$$

wherein $G_0^{(i)}$, $G_1^{(i)}$ and $G_2^{(i)}$ are expressed by appropriate dependencies for $t = t_i$.

The value t_i in Eqs. (9.18), (9.19), (9.20) and (9.21) is $t_i = (L_i - L_P^{(next)})/(L_K^{(next)} - L_P^{(next)})$, where L_i, $L_P^{(next)}$, $L_K^{(next)}$ is the mileage of points i, P and K.

Processing of the approximation equations to determine the values $\tan u_P$, $\tan u_K$ and $\tan\alpha$ is carried out according to the same rules, as described in Sect. 9.1.1 for the curves (9.1) and (9.2). Just as in that case, after determining $\tan u_P$ and $\tan u_K$ are required adequate controls, which relate to:

- fulfillment of the limit values by the determined inclinations $\tan u_P$ and $\tan u_K$,
- fulfillment of the limit values by the minimum radii of curvature R_E within the subsequent curves.

The controlled radii R_E result from a general Eq. (7.2). For the particular curves, it has a following form:

- for curves GTC_GS-1

$$R_E = x_K \frac{\left[1 + \left(N_1'(t_E) \tan u_P + N_2'(t_E) \tan u_K\right)^2\right]^{3/2}}{\left|N_1''(t_E) \tan u_P + N_2''(t_E) \tan u_K\right|} \qquad (9.22)$$

wherein $N_1'(t_E)$, $N_2'(t_E)$, $N_1''(t_E)$ and $N_2''(t_E)$ are expressed for the calculated value $t = t_E$, where

$$N_1' = 1 - 3t^2 + 2t^3$$

$$N_2' = 3t^2 - 2t^3$$

and

$$N_1'' = -6t + 6t^2$$

$$N_2'' = 6t - 6t^2$$

- for curves GSSTC_GS-2

$$R_E = x_K \frac{\left[1 + \left(M_0'(t_M)\tan\alpha + M_1'(t_M)\tan u_P + M_2'(t_M)\tan u_K\right)^2\right]^{3/2}}{\left|M_0''(t_M)\tan\alpha + M_1''(t_M)\tan u_P + M_2''(t_M)\tan u_K\right|} \qquad (9.23)$$

wherein $M_0'(t_E)$, $M_1'(t_E)$, $M_2'(t_E)$, $M_0''(t_E)$, $M_1''(t_E)$ and $M_2''(t_E)$ are expressed for calculated value $t = t_E$, where

$$M_0' = 30t^2 - 60t^3 + 30t^4$$

$$M_1' = 1 - 18t^2 + 32t^3 - 15t^4$$

$$M_2' = -12t^2 + 28t^3 - 15t^4$$

and

$$M_0'' = 60t - 180t^2 + 120t^3$$

$$M_1'' = -36t + 96t^2 - 60t^3$$

$$M_2'' = -24t + 84t^2 - 60t^3$$

- for curves GTC_GS-3

$$R_E = x_K \frac{\left[1 + \left(F_1'(t_E)\tan u_P + F_2'(t_E)\tan u_K\right)^2\right]^{3/2}}{\left|F_1''(t_E)\tan u_P + F_2''(t_E)\tan u_K\right|} \qquad (9.24)$$

wherein $F_1'(t_E)$, $F_2'(t_E)$, $F_1''(t_E)$ and $F_2''(t_E)$ are expressed for calculated value $t = t_E$, where:

$$F_1' = 1 - 10t^3 + 15t^4 - 6t^5$$

$$F_2 = 10t^3 - 15t^4 + 6t^5$$

and

$$F_1'' = -30t^2 + 60t^3 - 30t^4$$

$$F_2'' = 30t^2 - 60t^3 + 30t^4$$

- for curves GSSTC_GS-4

$$R_E = x_K \frac{\left[1 + \left(G_0'(t_M)\tan\alpha + G_1'(t_M)\tan u_P + G_2'(t_M)\tan u_K\right)^2\right]^{3/2}}{\left|G_0''(t_M)\tan\alpha + G_1''(t_M)\tan u_P + G_2''(t_M)\tan u_K\right|} \tag{9.25}$$

wherein $G_0'(t_E)$, $G_1'(t_E)$, $G_2'(t_E)$, $G_0''(t_E)$, $G_1''(t_E)$ and $G_2''(t_E)$ are expressed for calculated value $t = t_E$, where

$$G_0' = 140t^3 - 420t^4 + 420t^5 - 140t^6$$

$$G_1' = 1 - 80t^3 + 225t^4 - 216t^5 + 70t^6$$

$$G_2' = -60t^3 + 195t^4 - 204t^5 + 70t^6$$

and

$$G_0'' = 420t^2 - 1680t^3 + 2100t^4 - 840t^5$$

$$G_1'' = -240t^2 + 900t^3 - 1080t^4 + 420t^5$$

$$G_2'' = -180t^2 + 780t^3 - 1020t^4 + 420t^5$$

A value t_E that describes a location of curvature maximum must be first determined in order to control radii R_E. The value t_E result from solution of a nonlinear Eq. (7.4) after appropriate expressing of derivatives y', y'' and y'''.

The derivatives y', y'' and y''' have the following forms:

- for curves GTC_GS-1

$$y' = N_1' \tan u_P + N_2' \tan u_K \tag{9.26}$$

$$y'' = \frac{1}{x_K}\left(N_1'' \tan u_P + N_2'' \tan u_K\right) \tag{9.27}$$

$$y''' = \frac{1}{x_K^2}\left(N_1''' \tan u_P + N_2''' \tan u_K\right) \tag{9.28}$$

where

$$N_1''' = -6 + 12t$$

$$N_2''' = 6 - 12t$$

(formulas for N_1', N_2', N_1'', N_2'' are listed by Eq. (9.22)
- for curves GSSTC_GS-2

$$y' = M_0' \tan \alpha + M_1' \tan u_P + M_2' \tan u_K \tag{9.29}$$

$$y'' = \frac{1}{x_K} \left(M_0'' \tan \alpha + M_1'' \tan u_P + M_2'' \tan u_K \right) \tag{9.30}$$

$$y''' = \frac{1}{x_K^2} \left(M_0''' \tan \alpha + M_1''' \tan u_P + M_2''' \tan u_K \right) \tag{9.31}$$

where

$$M_0''' = 60 - 360t + 360t^2$$

$$M_1''' = -36 + 192t - 180t^2$$

$$M_2''' = -24 + 168t - 180t^2$$

(formulas for M_0', M_1', M_2', M_0'', M_1'', M_2'' are listed by Eq. (9.23)
- for curves GTC_GS-3

$$y' = F_1' \tan u_P + F_2' \tan u_K \tag{9.32}$$

$$y'' = \frac{1}{x_K} \left(F_1'' \tan u_P + F_2'' \tan u_K \right) \tag{9.33}$$

$$y''' = \frac{1}{x_K^2} \left(F_1''' \tan u_P + F_2''' \tan u_K \right) \tag{9.34}$$

where

$$F_1''' = -60t + 180t^2 - 120t^3$$

$$F_2''' = 60t - 180t^2 + 120t^3$$

(formulas for F_1', F_2', F_1'', F_2'' are listed by Eq. (9.24)

- for curves GSSTC_GS-4

$$y' = G_0' \tan \alpha + G_1' \tan u_P + G_2' \tan u_K \tag{9.35}$$

$$y'' = \frac{1}{x_K} \left(G_0'' \tan \alpha + G_1'' \tan u_P + G_2'' \tan u_K \right) \tag{9.36}$$

$$y''' = \frac{1}{x_K^2} \left(G_0''' \tan \alpha + G_1''' \tan u_P + G_2''' \tan u_K \right) \tag{9.37}$$

where

$$G_0''' = 840t - 5040t^2 + 8400t^3 - 4200t^4$$

$$G_1''' = -480t + 2700t^2 - 4320t^3 + 2100t^4$$

$$G_2''' = -360t + 2340t^2 - 4080t^3 + 2100t^4$$

(formulas for G_0', G_1', G_2', G_0'', G_1'', G_2'' are listed by Eq. (9.25).

9.2 Designing Vertical Arcs Using Polynomial Transition Curves

The polynomial transition curves can also be used to design of vertical arcs. Traditionally, the vertical arcs are formed using a quadratic parabola. The research described in the article (Kobryń 2014) showed that the arcs formed by the vertical transition curves (Fig. 9.8) make it possible to reduce fuel consumption in comparison with traditional parabolic arcs. This results from a reduction in average gradients within the vertical arc, if it is formed by the transition curves. Another advantage of some transition curves is the possibility to better match the arc to the vertical landform.

9.2.1 Designing Vertical Arcs Using Transition Curves with Horizontal Tangent at End Point

Procedures for designing of vertical arcs using the curves given by Eqs. (9.1) and (9.2) were described in article (Kobryń 2016). Starting point are two successive straight sections with inclinations s_I and s_{II}. The collapse of grade line between the two straight lines will be radiused using the transition curves given by Eq. (9.1) or (9.2). The initial point of the arc will be located at the beginning of the local

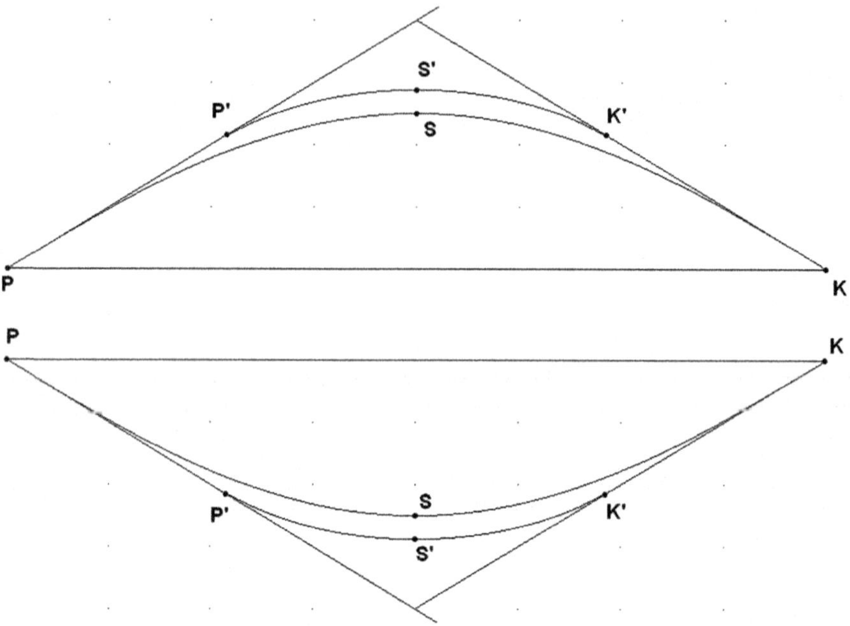

Fig. 9.8 Two types of vertical arcs (PSK—transition curves, PP'S'K'K—straight lines with quadratic parabola)

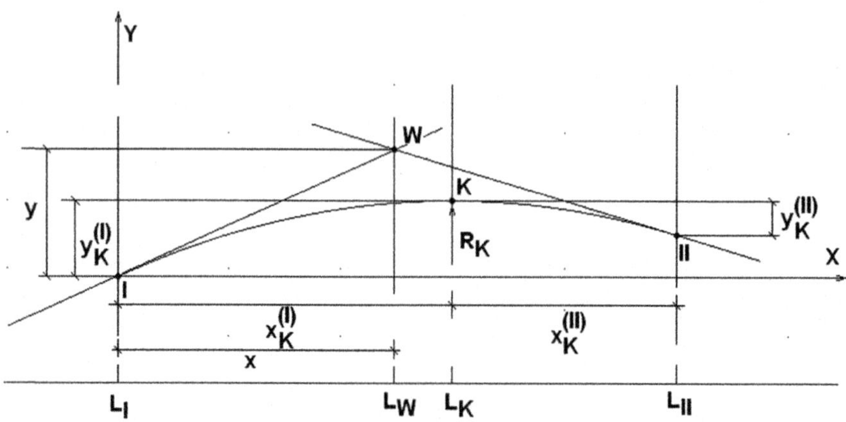

Fig. 9.9 Convex arc formed by the curves (9.1) or (9.2)

coordinate system of the designed vertical curve. First will be analyzed convex arc defined by the points I, II and K (Fig. 9.9), whereby the unknown location of the points I, II and K will be calculated according to the procedure described below.

It was assumed that the points I and II are appropriately a start points of first and second curve. It can be written:

$$\tan u_P^{(I)} = s_I \tag{9.39}$$

$$\tan u_P^{(II)} = -s_{II} \tag{9.40}$$

A location of the points I and II relative to the point W, however, is unknown and therefore needs to be determined. This is necessary to design a vertical curve. Assuming the same parameter E of the two curves and equal radius of curvature at the connecting point K, on the basis of Eq. (9.17) it can be written:

$$\frac{x_K^{(I)}}{\tan_P^{(I)}} = \frac{x_K^{(II)}}{\tan_P^{(II)}} \tag{9.41}$$

In the coordinate system as shown in Fig. 9.9, the first tangent equation has a following form:

$$y = \tan u_P^{(I)} \cdot x \tag{9.42}$$

Whereas, the second tangent equation has the form:

$$y = -\tan u_P^{(II)} \cdot (x - x_{II}) + y_{II} \tag{9.43}$$

where x_{II} and y_{II} are the coordinates of the point II in the coordinate system as show in Fig. 9.9. On the basis of Eqs. (9.42) and (9.43), for the point of intersection (W) of two tangents follows:

$$x\left(\tan u_P^{(I)} + \tan u_P^{(II)}\right) = y_{II} + x_{II} \tan u_P^{(II)} \tag{9.44}$$

However, the coordinates x_{II} and y_{II} of the point II can be expressed as:

$$x_{II} = x_K^{(I)} + x_K^{(II)} \tag{9.45}$$

$$y_{II} = y_K^{(I)} - y_K^{(II)} \tag{9.46}$$

whereby:

$x_K^{(I)}, x_K^{(II)}$ abscissas of the end points of the curve I and II in the coordinate system as shown in Fig. 9.9,

$y_K^{(I)}, y_K^{(II)}$ ordinates of the end point of the curve I and II, calculated on the basis of Eqs. (9.1) or (9.2).

Because for the point K is $x = x_K$, it follows $t = 1$. The values $y_K^{(I)}$ and $y_K^{(II)}$ can be expressed in the general form as:

- for curves (9.1)

$$y_K = \frac{6E+1}{12E} x_K \tan u_P \tag{9.47}$$

- for curves (9.2)

$$y_K = \frac{5E+1}{10E} x_K \tan u_P \tag{9.48}$$

Since, according to Eq. (9.41) is

$$x_K^{(II)} = x_K^{(I)} \frac{\tan u_P^{(II)}}{\tan u_P^{(I)}} \tag{9.49}$$

thus, on the basis of Eq. (9.45) follows:

$$x_{II} = x_K^{(I)} \frac{\tan u_P^{(I)} + \tan u_P^{(II)}}{\tan u_P^{(I)}} \tag{9.50}$$

In the case where the arc is formed by two vertical curves (9.1), on the basis of Eq. (9.46) and after appropriate transformations, we obtain the ordinate y_{II}:

$$y_{II} = x_K^{(I)} \frac{6E+1}{12E} \frac{\left(\tan u_P^{(I)} - \tan u_P^{(II)}\right)\left(\tan u_P^{(I)} + \tan u_P^{(II)}\right)}{\tan u_P^{(I)}} \tag{9.51}$$

Finally, on the basis of Eqs. (9.44), (9.50) and (9.51) it results:

$$x = x_K^{(I)} \frac{6E+1}{12E} \frac{\tan u_P^{(I)} - \tan u_P^{(II)}}{\tan u_P^{(I)}} + x_K^{(I)} \frac{6E+1}{12E} \frac{\tan u_P^{(II)}}{\tan u_P^{(I)}} \tag{9.52}$$

According to Fig. 9.9, the radius of curvature at the connecting point of the two curves is R. From Eq. (9.17) it follows that:

$$x_K^{(I)} = \frac{R_K \tan u_P^{(I)}}{E} \tag{9.53}$$

Taking into account Eq. (9.53) in (9.52), after appropriate transformation is obtained:

$$x = R_K \frac{(6E+1)\tan u_P^{(I)} + (6E-1)\tan u_P^{(II)}}{12E^2} \qquad (9.54)$$

This formula describes an abscissa of the point W (point of tangents intersection) in the local coordinate system as show in Fig. 9.9. On the basis of Eqs. (9.54) and (9.42) the ordinate y of the W is:

$$y = R_K \frac{(6E+1)\mathrm{tg}\, u_P^{(I)} + (6E-1)\mathrm{tg}\, u_P^{(II)}}{12E^2} \mathrm{tg}\, u_P^{(I)} \qquad (9.55)$$

In the case of the curves (9.2) is obtained, respectively:

$$x = R_K \frac{(5E+1)\tan u_P^{(I)} + (5E-1)\tan u_P^{(II)}}{10E^2} \qquad (9.56)$$

$$y = R_K \frac{(5E+1)\tan u_P^{(I)} + (5E-1)\tan u_P^{(II)}}{10E^2} \tan u_P^{(I)} \qquad (9.57)$$

As a result, the predetermined position of the point W, the values x and y defined by Eqs. (9.54) and (9.55) or (9.56) and (9.57) allow to determine the position of the point I and II. A mileage L_I and height z_I of the point I are:

$$L_I = L_W - x \qquad (9.58)$$

$$z_I = z_W - y \qquad (9.59)$$

In the case of end of the arc (the point II) the mileage and height are:

$$L_{II} = L_W - x + x_K^{(I)} + x_K^{(II)} \qquad (9.60)$$

$$z_{II} = z_W - y + y_K^{(I)} - y_K^{(II)} \qquad (9.61)$$

For the intermediate points located on the first part of arc (between the points I and K) is obtained:

$$L_i^{(I)} = L_W - x + x_i^{(I)} \qquad (9.62)$$

$$z_i^{(I)} = z_W - y + y_i^{(I)} \qquad (9.63)$$

where $y_i^{(I)}$ is determined by equation of the curve for $x_i^{(I)} = L_i - L_I$.

For the intermediate points located on the second part of arc (between the points K and II) is obtained:

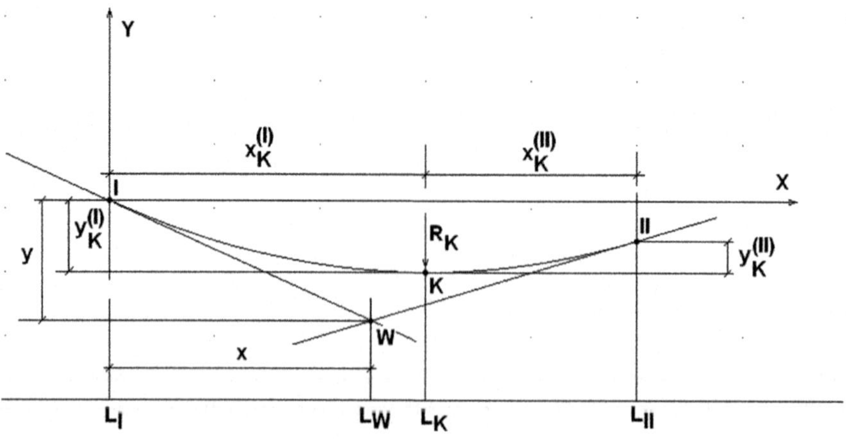

Fig. 9.10 Concave arc formed by the curves (9.1) or (9.2)

$$L_i^{(II)} = L_W - x + x_K^{(I)} + x_K^{(II)} - x_i^{(II)} \tag{9.64}$$

$$z_i^{(II)} = z_W - y + y_K^{(I)} - y_K^{(II)} + y_i^{(II)} \tag{9.65}$$

where $y_i^{(II)}$ is described by equation of the curve for $x_i^{(II)} = L_{II} - L_i$.

Similarly is performed a design procedure for the concave arc (Fig. 9.10). Based on the inclinations of the grade line designed as a series of straight lines it can be written:

$$\tan u_P^{(I)} = -s_I \tag{9.66}$$

$$\tan u_P^{(II)} = s_{II} \tag{9.67}$$

The equations of both tangents in the local coordinate system as shown in Fig. 9.10 have the form:

- in the case of the first tangent

$$y = -\tan u_P^{(I)} \cdot x \tag{9.68}$$

- in the case of the second tangent

$$y = \tan u_P^{(II)} \cdot (x - x_{II}) + y_{II} \tag{9.69}$$

A similar analysis as for the convex arc shows that the abscissa x of the tangents intersection is described by an equation analogous to (9.54) or (9.56). The ordinate y of the point W results from Eq. (9.68) as:

- for curves (9.1)

$$y = -R_K \frac{(6E+1)\tan u_P^{(I)} + (6E-1)\tan u_P^{(II)}}{12E^2} \tan u_P^{(I)} \qquad (9.70)$$

- for curves (9.2)

$$y = -R_K \frac{(5E+1)\tan u_P^{(I)} + (5E-1)\tan u_P^{(II)}}{10E^2} \tan u_P^{(I)} \qquad (9.71)$$

As a result, the unknown position of the point I can be described by Eq. (9.58) and equation:

$$z_I = z_W + y \qquad (9.72)$$

in which the value of y results from Eq. (9.70) or (9.71).

The position of the point II is described by Eq. (9.60) and equation:

$$z_{II} = z_W + y - y_K^{(I)} + y_K^{(II)} \qquad (9.73)$$

For intermediate points, it results:

- part of the arc between the points I and K

$$L_i^{(I)} = L_W - x + x_i^{(I)} \qquad (9.74)$$

$$z_i^{(I)} = z_W + y - y_i^{(I)} \qquad (9.75)$$

- part of the arc between the points K and II

$$L_i^{(II)} = L_W - x + x_K^{(I)} + x_K^{(II)} - x_i^{(II)} \qquad (9.76)$$

$$z_i^{(II)} = z_W + y - y_K^{(I)} + y_K^{(II)} - y_i^{(II)} \qquad (9.77)$$

whereby the values of $y_i^{(I)}$ and $y_i^{(II)}$ are determined on the basis of curve equation for $x_i^{(I)} = L_i - L_I$ and $x_i^{(II)} = L_{II} - L_i$.

9.2.2 *Designing Vertical Arcs Using General Transition Curves*

To designing vertical curves can be used also general transition curves, which polynomial solutions are presented in Sect. 7.2. This includes the curves marked with the following identification codes:

- curves GTC_GS-1

$$y = x_K \left(N_1 \tan u_p + N_2 \tan u_K \right) \tag{9.78}$$

where:

$$N_1 = t - t^3 + \frac{1}{2}t^4$$

$$N_2 = t^3 - \frac{1}{2}t^4$$

- curves GSSTC_GS-2

$$y = x_K \left(M_0 \tan \alpha + M_1 \tan u_P + M_2 \tan u_K \right) \tag{9.79}$$

where:

$$M_0 = 10t^3 - 15t^4 + 6t^5$$

$$M_1 = t - 6t^3 + 8t^4 - 3t^5$$

$$M_2 = -4t^3 + 7t^4 - 3t^5$$

whereby the values $\tan \alpha$, $\tan u_P$ and $\tan u_K$ should fulfill a condition:

$$\tan \alpha = M_{1/0} \tan u_P + M_{2/0} \tan u_K \tag{9.80}$$

where: $M_{1/0} \in \langle 2/5; 3/5 \rangle$ and $M_{2/0} \in \langle 2/5 ; 3/5 \rangle$
- curves GTC_GS-3

$$y = x_K \left(F_1 \tan u_p + F_2 \tan u_K \right) \tag{9.81}$$

where:

$$F_1 = t - \frac{5}{2}t^4 + 3t^5 - t^6$$

$$F_2 = \frac{5}{2}t^4 - 3t^5 + t^6$$

- curves GSSTC_GS-4

$$y = x_K(G_0 \tan \alpha + G_1 \tan u_P + G_2 \tan u_K) \tag{9.82}$$

where:

$$G_0 = 35t^4 - 84t^5 + 70t^6 - 20t^7$$

$$G_1 = t - 20t^4 + 45t^5 - 36t^6 + 10t^7$$

$$G_2 = -15t^4 + 39t^5 - 34t^6 + 10t^7$$

whereby the values $\tan \alpha$, $\tan u_P$ and $\tan u_K$ should fulfill a condition

$$\tan \alpha = G_{1/0} \tan u_P + G_{2/0} \tan u_K \tag{9.83}$$

wherein $G_{1/0} \in \langle 3/7\,; 4/7 \rangle$ and $G_{2/0} \in \langle 3/7\,; 4/7 \rangle$.

In the case of the general transition curves, appropriate design procedures will be similar to those described in Sect. 9.2.1. A main difference lies in the fact that this time the whole vertical arc is created only by one curve.

It is essential to determine the position of the zero point of the local coordinate system relative to the point W of tangents intersection. According to Figs. 9.11 and 9.12, can be created a following equations of straight lines which pass through the point W:

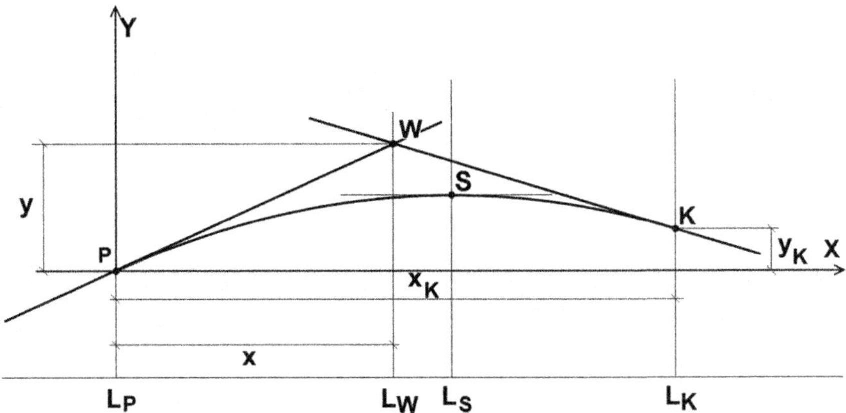

Fig. 9.11 Convex arc formed by general transition curve

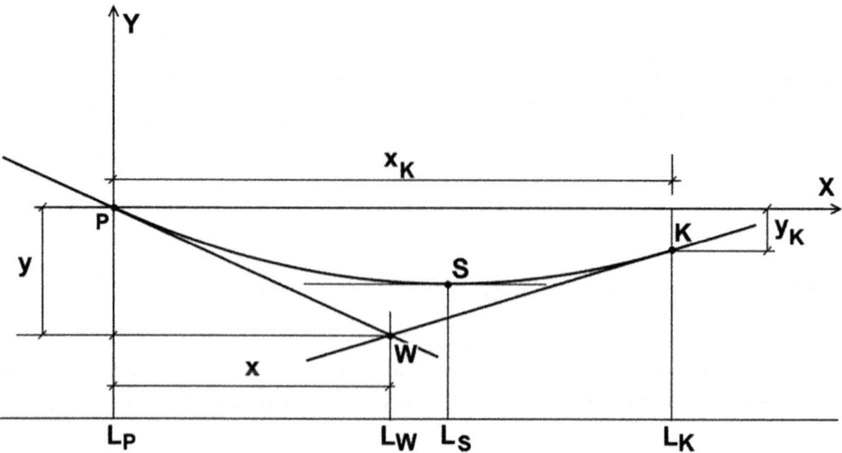

Fig. 9.12 Concave arc formed by general transition curve

$$y = \tan u_P \cdot x \tag{9.84}$$

$$y = \tan u_P \cdot (x - x_K) + y_K \tag{9.85}$$

From Eqs. (9.84) and (9.85) it follows:

$$x = \frac{x_K \tan u_K - y_K}{\tan u_K - \tan u_P} \tag{9.86}$$

whereby y_K is the ordinate of the K described by the use of appropriate Eqs. (9.78), (9.79), (9.81) or (9.83) for $t = 1$.

From Eq. (9.86) it follows that to determine the position of the point P must be know the length of the abscissa x_K. The abscissa x_K depends on the designed minimum radius $R_{\min}$. The value x_K can be calculated on the basis of Eq. (7.2). It follows from this:

- for curves (9.78):

$$x_K = R_{\min} \frac{\left| N_1''(t_E) \tan u_P + N_2''(t_E) \tan u_K \right|}{\left[1 + \left(N_1'(t_E) \tan u_P + N_2'(t_E) \tan u_K \right)^2 \right]^{3/2}} \tag{9.87}$$

- for curves (9.79):

$$x_K = R_{\min} \frac{\left| M_0''(t_E) \tan u_P + M_1''(t_E) \tan u_P + M_2''(t_E) \tan u_K \right|}{\left[1 + \left(M_0'(t_E) \tan u_P + M_1'(t_E) \tan u_P + M_2'(t_E) \tan u_K \right)^2 \right]^{3/2}} \tag{9.88}$$

- for curves (9.81):

$$x_K = R_{\min} \frac{\left| F_1''(t_E) \tan u_P + F_2''(t_E) \tan u_K \right|}{\left[1 + \left(F_1'(t_E) \tan u_P + F_2'(t_E) \tan u_K \right)^2 \right]^{3/2}} \tag{9.89}$$

- for curves (9.83):

$$x_K = R_{\min} \frac{\left| G_0''(t_E) \tan u_P + G_1''(t_E) \tan u_P + G_2''(t_E) \tan u_K \right|}{\left[1 + \left(G_0'(t_E) \tan u_P + G_1'(t_E) \tan u_P + G_2'(t_E) \tan u_K \right)^2 \right]^{3/2}} \tag{9.90}$$

In the above equations, the value of t_E is the location of the point where the curvature assumes a maximum value $k(t_E) = k_{\max} = 1/R_{\min}$. The value t_E can be determined on the basis of necessary condition for existence of the curvature extremum that is described by Eq. (7.4), whereby the equations expressed the derivatives y', y'' and y''' are given in Sect. 9.1.2.

It should be noted that the value x given by Eq. (9.86) is abscissa of the point W in the local coordinate system, as shown in Figs. 9.11 and 9.12. This value is equal to the difference of mileage of the points W and P, that is $x = x_W = L_W - L_P$. After determining of the value x_W, can be calculated the mileage and the height of the point P:

$$L_P = L_W - x_W \tag{9.91}$$

$$H_P = H_W - x_W \tan u_P \tag{9.92}$$

After determining of the values L_P and H_P, can be calculated position of the intermediate points, For this purpose can be used appropriate Eqs. (9.78), (9.79), (9.81) or (9.83).

References

Kobryń A (2002) Wielomianowe krzywe przejściowe w projektowaniu niwelety tras drogowych. Wydawnictwa Politechniki Białostockiej, Rozprawy Naukowe nr 100, Białystok (in Polish)

Kobryń A (2014) Transition curves in vertical alignment as a method for reducing fuel consumption. Baltic J Road Bridge Eng 9(4):260–268

Kobryń A (2016) Vertical arcs design using polynomial transition curves. KSCE J Civil Eng 20 (1):376–384

Kobryń A (2017) Optimizing of vertical alignment using general transition curves. KSCE J Civil Engineering, ##(#):##–##. (in review)

Printed by Printforce, the Netherlands